Metric-Driven Design Verification

An Engineer's and Executive's Guide to First Pass Success

Hamilton B. Carter
Shankar Hemmady

Metric-Driven Design Verification
An Engineer's and Executive's Guide to First Pass Success

Springer

Hamilton B. Carter
Cadence Design Systems, Inc.
San Jose, CA
USA

Shankar Hemmady
Cadence Design Systems, Inc.
San Jose, CA
USA

Library of Congress Control Number: 2007924215

ISBN 0-387-38151-1
ISBN 978-0-387-38151-0

e-ISBN 0-387-38152-X
e-ISBN 978-0-387-38152-7

Printed on acid-free paper.

© 2007 Springer Science+Business Media, LLC
All rights reserved. This work may not be translated or copied in whole or in part without the written permission of the publisher (Springer Science+Business Media, LLC, 233 Spring Street, New York, NY 10013, USA), except for brief excerpts in connection with reviews or scholarly analysis. Use in connection with any form of information storage and retrieval, electronic adaptation, computer software, or by similar or dissimilar methodology now know or hereafter developed is forbidden. The use in this publication of trade names, trademarks, service marks and similar terms, even if they are not identified as such, is not to be taken as an expression of opinion as to whether or not they are subject to proprietary rights.

9 8 7 6 5 4 3 2 1

springer.com

Table of Contents

The Authors	xi
Dedications	xiii
Preface	xv
Introduction	xix
Contributing Authors in Order of Appearance	xxi

PART I ANALYZING AND DRIVING VERIFICATION: AN EXECUTIVE'S GUIDE

Chapter 1 The Verification Crisis	3
Chapter 2 Automated Metric-Driven Processes	**13**
Introduction	13
The Process Model	15
The Automated Metric-Driven Process Model	16
Project Management Using Metric-Driven Data	28
What Are Metrics For?	29
Tactical and Strategic Metrics	29
Summary	30
Chapter 3 Roles in a Verification Project	**31**
Introduction	31
The Executive	31
Marketing	33
Design Manager	34
Verification Manager	34
Verification Architect/Methodologist	35
Design/System Architect	36
Verification Engineer	37
Design Engineer	38
Regressions Coordinator	39
Debug Coordinator	39
Summary	40
Chapter 4 Overview of a Verification Project	**41**
Introduction	41
Summary	49

Chapter 5 Verification Technologies — 51
Introduction — 51
Metric-Driven Process Automation Tools — 52
Modeling and Architectural Exploration — 58
Assertion-Based Verification — 63
Simulation-Based Verification — 70
Mixed-Signal Verification — 73
Acceleration/Emulation-Based Verification — 75
Summary — 78

PART II MANAGING THE VERIFICATION PROCESS

Chapter 6 Verification Planning — 81
Introduction — 81
Chapter Overview — 83
Verification Planning — 86
Summary — 105

Chapter 7 Capturing Metrics — 107
Introduction — 107
The Universal Metrics Methodology — 109

Chapter 8 Regression Management — 113
Introduction — 113
Early Regression Management Tasks — 114
Regression Management — 114
Linking the Regression and Revision Management Systems — 115
Bring-Up Regressions — 116
Integration Regressions — 119
Design Quality Regressions — 121
Managing Regression Resources and Engineering Effectiveness — 122
Regression-Centric Metrics — 123
How Many Metrics Are Too Many? — 125
Summary — 127

Chapter 9 Revision Control and Change Integration — 129
Introduction — 129
The Benefits of Revision Control — 131
Metric-Driven Revision Control — 132
Summary — 139

Chapter 10 Debug — 141
Introduction — 141

Table of Contents   vii

Debug Metrics	144
Summary	153

PART III EXECUTING THE VERIFICATION PROCESS

Chapter 11 Coverage Metrics — 157
Introduction	157

Chapter 12 Modeling and Architectural Verification — 163
Introduction	163
How to Plan	164
Tracking to Closure	165
Reusing Architectural Verification Environments	165
Summary	166

Chapter 13 Assertion-Based Verification — 167
Introduction	167
How to Plan	170
Tracking to Closure	175
Opportunities for Reuse	177
Summary	179

Chapter 14 Dynamic Simulation-Based Verification — 181
Introduction	181
How to Plan	183
Taxonomy of Simulation-Based Verification	187
Tracking to Closure	191
Summary	196

Chapter 15 System Verification — 197
Introduction	197
Coverification Defined	199
Advancing SoC Verification	201
List of Challenges	202
ARM926 PrimeXsys Platform Design	205
Closing the Gap	207
DMA Diagnostic Program	208
Connecting the DMA Diagnostic to the Verification Environment	212
Connecting the Main() Function in C	215
Writing Stubs	216
Creating Sequences and Coverage	217
Conclusion	219
References	220

Chapter 16 Mixed Analog and Digital Verification   221
    Abstract   222
    Introduction   222
    Traditional Mixed-Signal Verification   223
    Verification Planning   225
    Analog Mixed-Signal Verification Kit   229
    Conclusion   233
    Reference   234

Chapter 17 Design for Test   235
    Introduction   236
    Motivation   238
    A Unified DFT Verification Methodology   239
    Planning   240
    Executing   241
    Automating   243
    Test Case   245
    Benefits   248
    Future Work   249
    Conclusions   249
    References   250

PART IV CASE STUDIES AND COMMENTARIES

Metric-Driven Design Verification: Why Is My Customer a Better Verification Engineer Than Me?   255
    Abstract   255
    Introduction   256
    Section 1: The Elusive Intended Functionality   257
    Section 2: The Ever-Shrinking Schedule   265
    Section 3: Writing a Metric-Driven Verification Plan   270
    Section 4: Implementing the Metric-Driven Verification Plan   274
    Conclusion   277

Metric-Driven Methodology Speeds the Verification of a Complex Network Processor   279
    The Task Looked to be Complex   280
    Discovering Project Predictability   281
    A Coverage-Driven Approach, a Metric-Driven Environment   282
    A New Level of Confidence   283

Developing a Coverage-Driven SoC Methodology   285
    Introduction   285
    Verification Background   286
    Current Verification Methodology   289

Table of Contents

Coverage and Checking	292
Results and Futures	293

From Panic-Driven to Plan-Driven Verification Managing the Transition — **297**

Verification of a Next-Generation Single-Chip Analog TV and Digital TV ASIC — **303**
- Abstract — 303
- Introduction — 304
- The Design — 305
- Verification Challenges — 306
- Addition of New Internal Buses — 307
- Module-Level Verification — 309
- Data Paths and Integration Verification — 309
- Management of Verification Process and Data — 309
- Key Enablers of the Solution — 310
- Results — 320
- Conclusions — 322
- Future Work — 322

Management IP: New Frontier Providing Value Enterprise-Wide — **325**

Adelante VD3204x Core, SubSystem, and SoC Verification — **329**
- Abstract — 330
- Introduction — 330
- Project Background — 331
- Verification Decisions — 333
- DSP Core Verification — 335
- DSP Subsystem Verification — 338
- SoC-Level Verification — 341
- Results and Future Work — 342

SystemC-based Virtual SoC: An Integrated System-Level and Block-Level Verification Approach from Simulation to Coemulation — **345**
- Abstract — 346
- Introduction: Verification and Validation Challenges — 347
- Virtual SoC TLM Platform — 348
- Functional Verification: Cosimulation TLM and RTL — 350
- Validation: Coemulation TLM-Palladium — 352
- Conclusion and Future Developments — 353

Is Your System-Level Project Benefiting from Collaboration or Headed to Chaos? — **355**

Index — **359**

The Authors

Hamilton Carter

Hamilton Carter has been awarded 14 patents in the field of functional verification. The patents address efficient sequencers for verification simulators, MESI cache coherency verification and component-based reusable verification systems. Carter worked on verification of the K5, K6, and K7 processors and their chipsets at AMD. He staffed and managed the first functional verification team at Cirrus Logic and has served as a manager, engineer, or consultant on over 20 commercial chips and EDA projects.

Shankar Hemmady

Shankar Hemmady is a senior manager at Cadence responsible for verification planning, methodology, and management solutions. Mr. Hemmady has verified and tested, or managed the functional closure of over 25 commercial chips over the past 18 years during his tenure in the industry as an engineer, manager, and consultant at 12 companies, including AMD, Cirrus Logic, Fujitsu, Hewlett Packard, Intel, S3, Sun, and Xerox.

Dedications

To my Parents who removed the word "cannot" from my vocabulary!

Hamilton Carter

To Seema, Shona, & Anand who make each and every moment a special one!

Shankar Hemmady

Preface

With the alarming number of first pass silicon functional failures, it has become necessary for all levels of engineering companies to understand the verification process. This book is organized to address all verification stakeholders at all levels of the engineering organization. The book is targeted at three somewhat distinct audiences:

- *Executives*. The people with their jobs on the line for increasing shareholder value.
- *Project, design, and verification managers*. The people responsible for making sure each design goes out on time and perfect!
- *Verification and design engineers*. The innovators responsible for making sure that the project actually succeeds.

The book is divided into three parts corresponding to its three audiences. The level of technical depth increases as the book proceeds.

Part I gives an overview of the functional verification process. It also includes descriptions of the tools that are used in this flow and the people that enable it all. After outlining functional verification, Part I describes how the proper application of metric-driven techniques can enable more productive, more predictable and higher quality verification projects. Part I is targeted at the executive. It is designed to enable executives to ask appropriate educated questions to accurately measure and control the flow of a project.

Part I also holds value for project managers and verification engineers. It provides an overall view of the entire chip design process from a verification perspective. The chapters on a typical verification project and the overview of verification technologies will be of use to entry level verification engineers as well. This part of the book also provides a unique viewpoint on why management is asking for process data and how that data might be used.

Part II describes the various process flows used in verification. It delves into how these flows can be automated, and what metrics can be measured to accurately gauge the progress of each process. Part II is targeted at design and verification project managers. The emphasis is on how to use metrics within the context of standardized processes to react effectively to bumps in the project's execution.

Part III's audience is the design and verification engineering team. It focuses on the actual verification processes to be implemented and executed. This section of the book is divided with respect to the various verification technologies. Each chapter on a given technology is further subdivided into sections on how to plan effectively, and how to track metrics to closure.

Entire books have been written on implementing verification using the technologies discussed in Part III. We will not reiterate what those excellent volumes have already stated, nor do we intend to reinvent the wheel (yet, we are engineers after all). Implementation details will be discussed when they will make the metric-driven techniques discussed more effective.

Part IV contains various case studies and commentaries from experts in the metric-driven verification field.

The various parts of the book can also be described as a progression of process abstractions. The layers of abstractions are "Observational Processes," "Container Processes," and "Implementation Processes."

Observational Processes

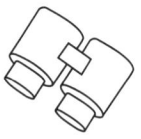
Part I looks at the verification process from an observational point of view. The various aspects of a project that should be observed are described to the reader along with informal suggestions about how to strategically manage a verification project based on these observations.

Container Processes

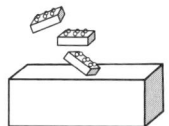
Part II looks at processes that are necessary regardless of the verification technology you are using; processes such as regression management, revision control, and debug. Part II describes how to implement these processes using metric-driven methodologies. It also also discusses the inter-relations of these processes.

Implementation Processes

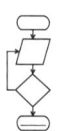
Part III describes each of the verification technologies and explores how a metric-driven methodology can be used to enhance the productivity, predictability, and quality offered by each of these technologies.

Finally, *Part IV* leaves the world of abstraction altogether and presents several concrete case studies that illustrate metric-driven processes in action. In addition to these case studies are several commentaries offered by industry experts in metric-driven methodologies.

Introduction

Legend has it that 2300 years ago, Euclid walked the beaches of Egypt with his students. They were exploring the fundamentals of a new field: geometry. Each day, Euclid would draw a new problem in the sandy shores of the Mediterranean Sea. He'd ask his students to reflect on each problem and discover what they could. One day he sketched a diagram that would come to be known as Euclid's 42nd Problem.

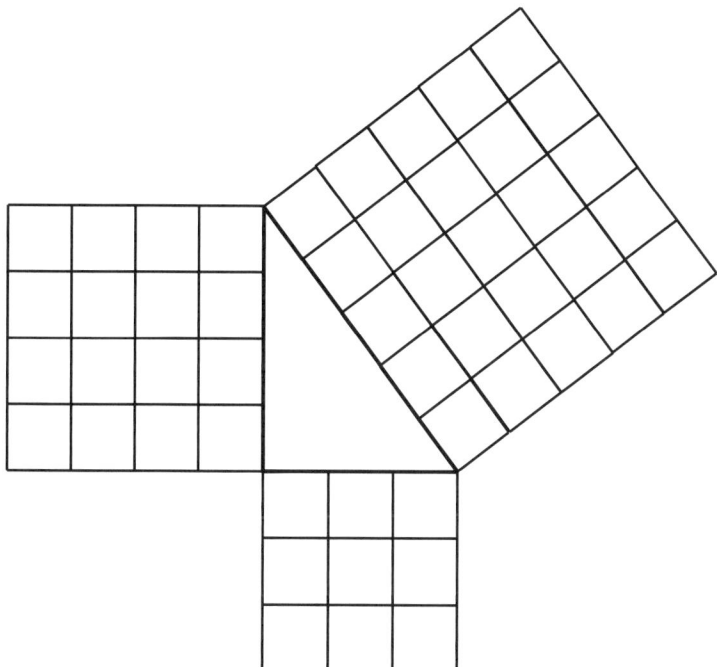

One of his particularly bright students worked on the diagram and came back with a simple formula:

$$a^2 + b^2 = c^2$$

This formula became so famous that it is now known simply by its discoverer's name: the Pythagorean Formula.

Pythagoras thirsted for knowledge and spent most of his life traveling the various countries of the ancient Hellenic world searching for it. In his travels, he encountered many cultures and gleaned valuable knowledge from each of them applying it to the burgeoning new field of geometry.

Today we're witnessing the birth of another new field, Metric-Driven Verification. Like Euclid, we hope to layout templates that not only illustrate the basics of this promising new field, but also inspire the reader to make even greater discoveries. Like Pythagoras, we have traveled the world searching for the best applications of this knowledge.

This book contains more than our basic understanding of the principles of metric-driven verification. The book also contains examples and experiences gleaned from many industry experts in verification and design. All of these are presented in their entirety in Part IV.

The last three chapters of Part III are about emerging technologies in the field of metric-driven verification:

- System verification
- Mixed-signal verification
- Verification of DFT hardware

These chapters use a different format. Each chapter contains a complete case study from one of the industry leaders in each of these three emerging areas.

Contributing Authors in Order of Appearance

Jason Andrews

Jason Andrews is a project leader at Cadence Design Systems, where he is responsible for hardware/software coverification and methodology for system-on-chip (SoC) verification. He is the author of the book "Co-Verification of Hardware and Software for ARM SoC Design" and holds a bachelor's degree in electrical engineering from The Citadel (Charleston, SC) and a master's degree in electrical engineering from the University of Minnesota (Minneapolis).

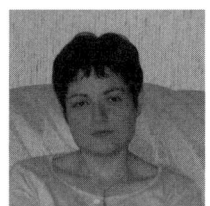

Monia Chiavacci

Ms. Chiavacci cofounded Yogitech in 2000. She is responsible for the mixed signal division. She worked as an analog designer from 1998 to 2000 after receiving her degree cum laude in electronic engineering at the Pisa University. Her work experiences include high-reliability systems in critical environments such as biomedical, space, and high-voltage automotive applications.

Gabriele Zarri

Mr. Zarri is a verification engineer at Yogitech. He is responsible for the development of verification IPs, verification environments for many international customers, and trainings on verification methodologies. His experience includes automotive protocols such as LIN, CAN, and Flexray. He is expert in OCP protocol, a universal complete socket standard for SoC design, and has recently acquired experience in the verification of mixed signal circuits. Gabriele specialized in Microelectronics and Telecommunications with an MS from Nice Sophia-Antipolis University.

Egidio Pescari

Egidio is a senior design and verification engineer at Yogitech. Prior to Yogitech, Mr. Pescari developed systems in critical environments such as automotive and space applications. He acquired experience in many automotive protocols such as LIN and CAN. He graduated from the University of Perugia in 1998.

Stylianos Diamantidis

Stylianos Diamantidis is the Managing Director of Globetech Solutions. Mr. Diamantidis is responsible for driving IP product strategy, engineering and consulting services. Prior to cofounding Globetech Solutions, he managed SGI's systems diagnostics group, spanning across servers, supercomputers, and high-end graphics product lines. His current areas of interest include advanced design verification methodologies, embedded systems, silicon test, debug, and diagnosis. Stylianos holds a B.Eng. from the University of Kent at Canterbury, UK and a MS in electrical engineering from Stanford University, USA. He is a member of the IET, IEEE, and IEEE-DASC.

Iraklis Diamantidis

Iraklis Diamantidis is a founder and senior verification engineer at Globetech Solutions. His current areas of interest include Electronic System-Level Design, Advanced Design Verification Methodologies, Silicon Test, Debug and Diagnosis, and System Software. Iraklis holds a B.Eng. from the University of Kent at Canterbury, UK, and a MS in electrical engineering from Stanford University. He is a member of the IET and the IEEE.

Thanasis Oikonomou

Thanasis Oikonomou is a senior digital systems designer and verification engineer at Globetech Solutions. His interests include computer architecture, high-speed networks, digital design, verification, and testing. He received BSc and MSc in computer science from the University of Crete, Greece.

Jean-Paul Lambrechts

Jean-Paul Lambrechts has over 20 years experience in leading hardware design in the networking and computer areas. His experience covers board-level hardware design, FPGA, and verification. Jean-Paul has now been with Cisco for 9 years where he has been responsible for line cards, packet forwarding engines, and layer 4–7 processor card. Jean-Paul holds a MSEE degree from the Louvain University in Belgium.

Alfonso Íñiguez

Alfonso Íñiguez is a principal staff verification engineer with the Security Technology Center at Freescale Semiconductor, where he is the verification lead responsible for developing, improving and applying functional verification tools, and methodologies. His work includes cryptographic hardware accelerator design. He holds a B.S. in computer engineering from the Universidad Autónoma de Guadalajara, México, and an MS in electrical engineering from the University of Arizona.

Dr. Andreas Dieckmann

In 1995, after obtaining his MA at the University of Erlangen and his Ph.D. in electronic engineering at Technical University of Munich, Dr. Dieckmann began working at Siemens AG. He was initially responsible for board and fault simulation. From 1997, Dr. Dieckmann gained expertise in system simulation and verification of ASICs. Since 2001, he has been in charge of coordinating and leading several verification projects employing simulation with VHDL and Specman "e," formal property and equivalence checking, emulation, and prototyping.

Susan Peterson

Susan Peterson has been trying to escape from the EDA industry for the past 20 years, where she has spent her time listening to customers and trying to help them to solve their critical problems in various sales and marketing roles. Prior to that, she was a practicing engineer, and earned her MBA from the University of Denver.

Paul Carzola

Paul Carzola is a senior consulting engineer for verification at Cadence. He received a Bachelor of Science Degree in computer engineering at Florida Atlantic University in 1995. Since then, Paul has spent the last 10 years in functional verification and the pursuit to finding effective and powerful methods to verification while making it easier and enjoyable to apply. For the past 5 years, he has served in a consulting role in the area of functional verification methodology and has seen first hand the power of a Coverage-Driven approach.

Contributing Authors in Order of Appearance

YJ Patil

YJ Patil is a senior verification engineer at Genesis Microchip, where he is responsible for managing the verification of digital television (DTV) controller ASICs. Prior to Genesis, Mr. Patil was a verification engineer at several technology leaders including ATI, Silicon Access Networks, and Philips Semiconductors. He was a board designer at Tektronix. Mr. Patil holds an MS in software systems from BITS Pilani, India and BE in electronics and communication from Gulbarga University, India.

Dean D'Mello

Dean D'Mello is a solutions architect at Cadence Design Systems. He works closely with key customers worldwide to deploy advanced verification technologies, and with R&D to plan, develop, and introduce new methodologies and products. Prior to Cadence, Mr. D'Mello held ASIC design and verification roles at LSI Logic, Cogency Semiconductor, and Celestica, and product and test engineering roles at IBM. Dean holds a Masters of Applied Science (MASc) in electrical and computer engineering, from the University of Toronto, Canada.

Steve Brown

Steve Brown is Director of Marketing for Enterprise Verification Process Automation at Cadence Design Systems. He is a 20-year veteran of the EDA verification industry and has held various engineering and marketing positions at Cadence, Verisity, Synopsys, and Mentor Graphics. He specializes in solving engineering, management, and marketing challenges that arise when new technology and products enter the market. He earned BSEE and MSEE degrees from Oregon State University and has studied marketing strategy at Harvard, Stanford, Kellogg, and Wharton.

Roger Witlox

Roger Witlox joined Philips Research Laboratories in Eindhoven, The Netherlands in 1992, where he has been working on optical coherent communications systems and access networks. Mr. Witlox was earlier involved in the development of analog laser temperature and current control system. In 2000, he joined the CTO organization at Philips Semiconductors, where he was responsible for development and support of an in-house verification tool. He has been responsible for functional verification methodologies for hardware IP and was a member of the Verification Technical Work Group of the SPIRIT consortium. In 2004, he joined the DSP Innovation Center and is currently focusing on DSP subsystems, both specification and verification.

Ronald Heijmans

Ronald Heijmans studied at the Hoge School Eindhoven and graduated in 1992 in the field of "Technical Computer Science." He started his career as a PCB designer at the Philips Research Laboratories. Later, Mr. Heijmans focused on DSP algorithm design and applications for multichannel audio and speech coding. In 1999, he became a verification engineer at ESTC Philips Semiconductors, where he focused on DSP core and subsystems. Currently, as a verification architect, Ronald is defining a new environment including new verification methodologies.

Chris Wieckardt

Chris Wieckardt has been a verification engineer at Philips Semiconductors, Adelante Technologies and NXP Semiconductors in Eindhoven, The Netherlands since 2000. Prior to Philips, Mr. Wieckardt was a digital design engineer at Océ Research and Development, Venlo, The Netherlands.

Contributing Authors in Order of Appearance   xxvii

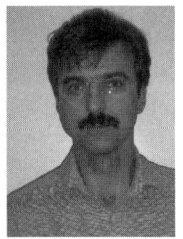

Dr. Laurent Ducousso

Laurent Ducousso has over 20 years of experience in digital design and verification. In 1994, Dr. Ducousso joined STMicroelectronics as the verification expert on CPU, microcontroller and DSP projects. Since 2000, he has managed the Home Entertainment Group verification team. Prior to STMicroelectronics, he was engaged in mainframe CPU development at Bull S.A for 8 years. Laurent holds a Ph.D. in computer sciences from Paris, France.

Frank Ghenassia

Frank Ghenassia is Director of the System Platforms Group in the HPC (Home, Portable, and Communication) sector at STMicroelectronics. Mr. Ghenassia focuses on IP/SOC verification, architecture definition, platform automation, and embedded software development based on high-level modeling approaches. He joined STMicroelectronics in 1995 and has worked on OS development, software debuggers, and system-to-RTL design flow activity. Mr. Ghenassia received his MS in electrical engineering in Israel.

Dr. Joseph Bulone

Joseph Bulone manages a team that provides central services in hardware emulation to STMicroelectronics divisions. Joseph defines and provides hardware-accelerated platforms for IP/SoC verification and software development. He joined the Central R&D division of STMicroelectronics in 1989, and was initially involved in the design of ATM chips. He began working on hardware emulation in 1993. He has been in charge of video chip validation, and hardware software co-design. He holds a Ph.D. in microelectronics from the Institut National Polytechnique de Grenoble, France.

Part I
Analyzing and Driving Verification: An Executive's Guide

Chapter 1
The Verification Crisis

If everything seems under control, you're not going fast enough.
– Mario Andretti

The time is at hand! This book proposes to revolutionize verification engineering! "It's rote work," you say? Can't be done!? Well get ready to be surprised and even mystified!

What is Verification?
So what is verification? Simply put, it is a process that ensures the *implemented device* will match the *product intent* defined for the device prior to sending the device for manufacturing. Notice the selection of words in the previous sentence. It didn't mention the device specification, or the device requirements. Every document that corresponds to the device (such as a specification or requirements list), is merely a translation of the actual intent of the device functionality as originally conceived. This is an important distinction. All the methodologies in this book will have at their heart, the goal of ensuring that the device does what it was *intended* to do, not necessarily what it was documented to do. Quite frequently, the first defects we find are specification issues, *not* design defects. Figure 1.1 shows the many translations of intent.

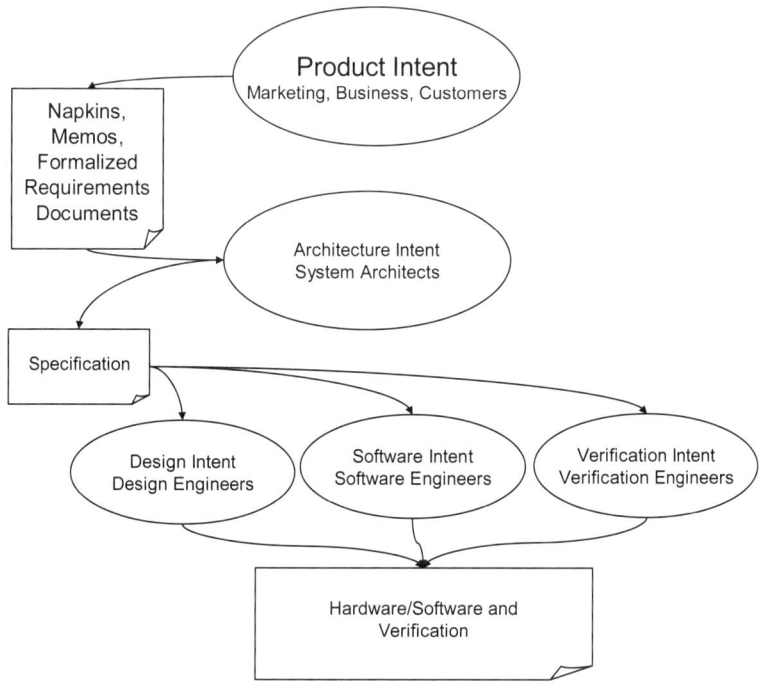

Figure 1.1 Intent Translation

The Crisis

The size of designs is increasing. Market window size is decreasing. These factors combine to create a rapidly increasing cost of failure (Table 1.1).

As designs become more and more complex and market windows become tighter and tighter, verification becomes crucial. More and more devices are now going directly into the mainstream consumer market. The mainstream consumer expects all features of a device to work properly. If they don't the consumer will return the device, get their money back, and go with a different supplier. There's really no room for error.

Rapidly shrinking silicon geometries have been both a blessing and a curse. It is possible to build more powerful, feature rich devices than ever before. However, along with all the new features comes an exploding multidimensional space for verification requirements.

1 The Verification Crisis

Table 1.1 Design Size, Market Window, and Cost of Failure

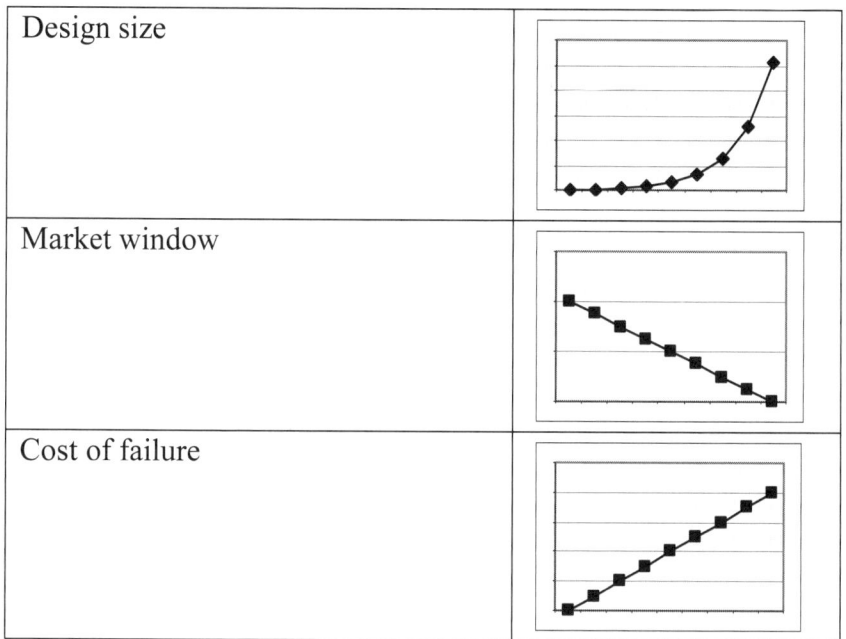

Design size	
Market window	
Cost of failure	

For example, consider a "simple" digital sound output port. The port can output sound in mono or stereo mode. In stereo mode, the sound frames can be transmitted with either the left channel or right channel first. Sound can be output in 8-, 12-, 16-, or 24-bit resolution. The gap between sound samples can be 0, 1, or 2 bits. In addition to all these specifications on the format of the output stream, the port can also be configured to use five different FIFO sizes for buffering input data and can run in either polling or interrupt-driven mode. This simple output port has over 240 functional combinations that must be verified. If even one of these combinations fails and it's the combination that our key customer had to have, we're facing a costly silicon respin.

Respins are expensive at more than a million dollars a piece, but in today's accelerated business atmosphere, there are even worse repercussions. A nonfunctional first tape-out can result in the loss of a job, the closing of a company and the ruin of a career. Clearly, it's important to get verification right the first time. By getting verification right the first time, companies can save millions of dollars on respins alone. Then they can make millions more by hitting their market windows on time.

The Need for Metric-Driven Processes
So, how do we solve the verification crisis? How do we ensure that our designs will go out "first silicon clean" every time? With a cultural change and newly available technology, it's actually quite simple.

For years verification has been done in a rather haphazard manner. Each company or project team within a company slowly assembled their own best practices. Some project teams developed very successful, rigorous processes for making sure verification was implemented and managed correctly. Others executed on their verification projects in a haphazard way. Still other teams did verification merely as an afterthought as the project started to wind up. The process-oriented teams had far higher success rates.

Effective project closure tracking was also frequently ignored. Here again, many disparate techniques have been documented and used. Some of these techniques included bug rate tracking, code coverage, functional coverage, and everyone's favorite: "Tape it out because management said so!"

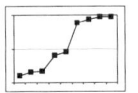 By objectively tracking important metrics, management can allocate resources more effectively, better predict the schedule of the project, and ensure a higher quality of the final product. Management and engineering productivity can be further enhanced if these objective metrics can be measured automatically. This book will show how to define what metrics are important to measure, how to measure those metrics automatically, and how to most effectively utilize those metrics to streamline engineering processes.

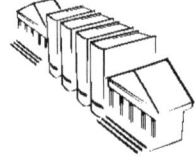 While other disciplines have reaped great rewards in productivity and effectiveness by moving to well-documented, accepted and established methodologies, ASIC design engineering is one of the few

1 The Verification Crisis

engineering activities where a "cowboy" mentality is still accepted and even expected! In other areas where large teams integrate work flows, processes have been defined for years. Accounting has the FASBs, manufacturing has ISO standards. No one argues about the format of a ledger entry, they worry about more important things like the actual analysis of the financial data. No one argues about where the header block on an architectural drawing should be placed or on the size of the page. They concentrate their effort on the actual architectural design.

Let's look at a small example of why tracking progress is so crucial to any activity. Imagine that on your rare Sunday off, you sit down in front of the TV, cold beer in hand and turn on your favorite sporting event. As the players enter the field, we hear the commentators begin to speak.

"Jack, someone will definitely win this game today. Both teams have entered the field with that goal in mind, and we feel it will definitely happen."

"Folks we're really not certain what two teams are playing today, but we've got someone looking into it and we'll have that information to you as soon as we can."

"After all, what is important is that the teams play often and hard, right? We expect to see lots of really hard effort put in today."

"Ah, and the players have begun. There's a really tall player (Fred get me his name), carrying the ball down the field. Oh! He's been tackled by a rather small chap (Fred, we're going to need another name!) And, the team is up and carrying the ball again! Did anyone think to find out how many yards to first down? Folks, we'll get you more stats right after this commercial break!"

When we watch sports, we want to know everything about the game from the first instant, right? We don't give the teams respect for beginning the game immediately and running around willy-nilly with the ball when we know nothing about the game, do we? Then why are we so content to execute on our engineering projects in this manner?

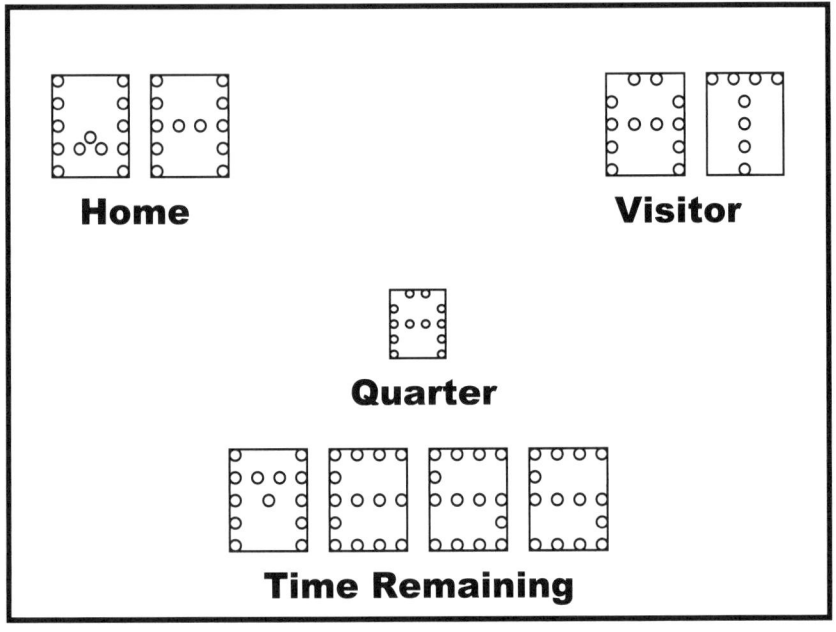

When we watch our sporting event, we expect to have a multitude of information at our fingertips:

- The amount of time left in the game
- The score of the game
- Progress toward the current goal
- The history and statistics of the player that most recently carried the ball

As the coach of the team, we'd expect to have all the information above and much more like:

- What to do when the opposing team does something we don't expect, like fumbling the ball
- The statistics of each of the players on the opposing team
- A plan for how to counter each of the other team's plays

1 The Verification Crisis

- Information about how our players match up vs. the players on the other team
- How each player on both teams is playing vs. their statistics

To accumulate this data, we'd employ an entire coaching staff to gather and analyze data both before the game and as the game progressed. Before the game, we'd build a plan of what we expected to do based on available data. As the game progressed we'd constantly adjust our plan to work with the situation at hand. And that's exactly how we should be executing our engineering projects.

But maybe we don't have to hire that pricey coaching staff. Maybe we can automate that part.

The message so far has been:

- Verification is hard! Brutally hard!
- If we're going to successfully verify today's designs we *have* to move to a process-oriented approach.
- Process isn't enough, we also have to be able to measure the output of our processes and use that information to adjust our direction.

Using emerging technology, we're going to show you how to move to a metric-driven, process-oriented verification flow. In Chapter 2, we'll outline exactly what these processes look like and how we measure and use process metrics. And don't worry, *we will* replace all those coaches with an automated system that will automatically capture and analyze metrics.

Now we'll spend a little bit of space explaining the logistics of the book so you can get the most out of it.

The Verification Hierarchy of Needs

In the year 1943, Maslow unveiled the hierarchy of needs to the world. This hierarchy described a set of basic needs that humans strive after. Each new level of needs can be reached only after the level before it has been attained. Figure 1.2 shows the verification hierarchy of needs.

Like Maslow's hierarchy, each additional level can only be fully attained and appreciated once the levels below it are realized.

The first level of the verification hierarchy is visibility. Visibility is paramount! Without it, the verification team is quite literally stumbling around in the dark. As we discussed above, without metrics that provide data about our engineering processes and visibility into those metrics, we're lost! Visible metrics make our schedules predictable and give us a measure of the quality of the device under verification.

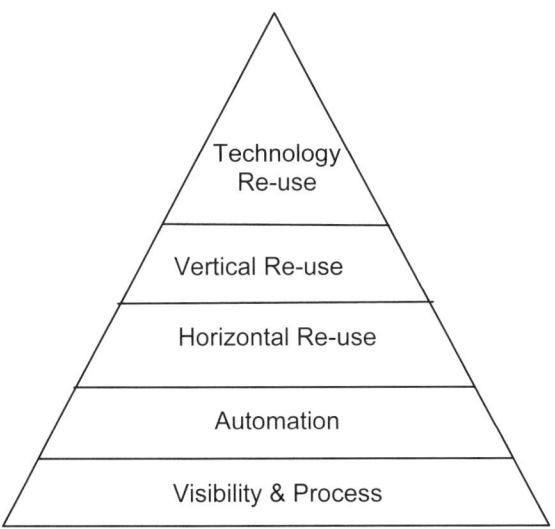

Figure 1.2 The Verification Hierarchy of Needs

The second level of the pyramid is automation. Once we have a handle on what we've planned to do, and how we will measure our execution, the next question is how do we do verification more productively? The first answer is to automate verification processes where possible. By automating verification processes, we increase the productivity of our engineers and free them up to tackle harder tasks. We also increase the predictability of the schedule by reducing the time required to complete the automated tasks. Finally, because our engineers have more time, they can improve the quality of the device by performing verification tasks that may have been left out of the schedule otherwise.

Visibility is required to automate our processes. We will show how the metrics themselves can be used to automate several tasks. Metrics are also required in some cases to gauge the effectiveness of automation. For example, without metrics, constrained random stimulus offers an effective method to explore the state space of a device for bugs that would not have been found otherwise. But, when we use metrics in the form of functional coverage in conjunction with constrained random stimulus, we have a much more powerful automation tool that not only explores our state space, but also automatically creates our testcases!

Once we have automated our processes, and we're no longer spending our mental effort doing rote work, we start to look at how we can reuse our creative work. That will be the subject of the next book in this series! First things first!

Chapter 2
Automated Metric-Driven Processes

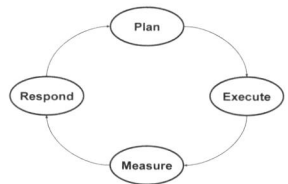

Introduction

Historically, many verification projects have been performed as an afterthought. They have been understaffed, under-planned, and under-executed. With today's complex design, it is widely agreed that verification consumes up to 70% of the total effort for a typical design project.

There are several issues that plague verification efforts. Among them are:

- *Insufficient planning.* High-priority issues are brought to light in the last stages of the project causing huge upheavals in resources and scheduling.
- *Lack of visibility.* Projects are frequently tracked by human updates. This is also known as "death by status meeting."
- *Scheduling issues.* Why is it a well-excepted axiom that the last 20% of the work will consume 80% of the available time? Shouldn't the last 20% of the effort take 20% of the time?
- *Inefficient use of tools.* The EDA industry has promoted verification solutions for years. It's well accepted that the verification effort required 70% of the total design cycle effort. Why hasn't this number changed in years in face of the advanced solutions available?

In this chapter, we'll outline the basis for a methodology that will resolve these issues. This methodology is based on automated metric-driven processes. The methodology is enabled by a new class of tools called metric-driven process automation tools, or MPA tools.

Processes are an important start to our solution. By using repeatable processes we can improve the predictability of our projects. A framework for modeling processes will be described.

But, processes aren't enough. Without visibility into the workings of these processes we are unable to track progress and respond to issues in an efficient manner. We'll describe how to identify metrics that should be tracked during the life of the process. These metrics will give us constant insight into the process's progress.

Even well-defined processes and metrics that track their progress aren't enough. The nature of the metrics is also important. The classic 20/80 situation described above is an example of a metric-driven process that doesn't work. In this case the metric is an engineer's opinion of the completeness of a given task. The tracking mechanism is a query from management. In order to be useful metrics must be objective rather than subjective and be capable of being automatically captured and tracked. *We have to remove human interpretation and reporting of metrics from the equation.*

MPA tools facilitate the methodology described. They facilitate the planning phase by enabling users to define what metrics will be used while planning. They can control our verification engines removing that tedious time consuming task. They automatically capture the metrics that are produced by the verification engines. Finally, they offer analysis engines that can process the metric data. The analysis engines can be used in conjunction with the execution control aspects of MPA tools to completely automate some processes.

Using automated metric-driven processes, we'll be able to better plan our work defining exactly what needs to be done in a manner that's measurable. These automated measurements will allow us to efficiently respond to issues as they arise. We'll even be able to use these metrics to further automate some of our manual processes and increase our operational efficiency.

Next we'll define the process model that will be used throughout the book.

The Process Model

For the purposes of this book, a process is any activity that can be modeled using the flow shown in Figure 2.1.

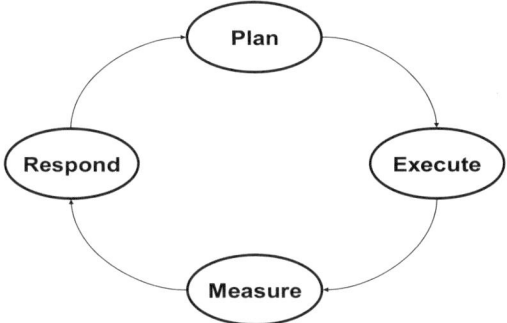

Figure 2.1 Verification Process Model

The flow consists of four phases. These are planning, execution, measurement, and response.

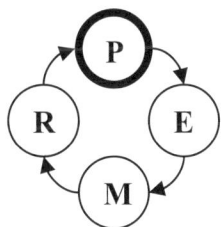

The first step of the process model is planning. This is where we determine what needs to be done and how to measure that it was in fact done. To efficiently execute a process, we need to know what the process hopes to achieve. Next, we need an objective way of knowing that the process has in fact achieved its goal.

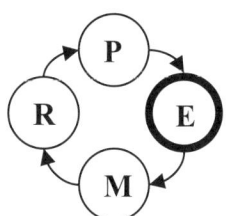

Once the process is planned, we need to make it happen. That brings us to the execution phase of the model. During the execution phase, we will act on our plans. Using our available human resources and verification engines, we'll create the verification environments we specified during planning.

These environments will create objective metric output that we'll use to gauge the completeness of the plan. We will use MPA tools to control the execution engines.

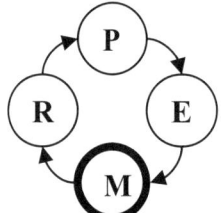

As our engines operate we need to measure the effectiveness of our efforts. During the planning phase we specified the metrics that will be used to gauge this effectiveness. The MPA tool will automatically gather the specified metrics from our execution engines. Some typical metrics include:

- RTL code coverage
- Functional coverage
- Assertion coverage
- Software code coverage
- Error messages and types
- Revision control information

The user specifies how these metrics are to be annotated back to the plan during the planning phase.

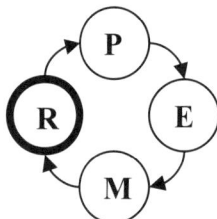

In the response phase, the user acts on the results of the data analysis performed during the measure phase. This analysis will be used to adjust existing plans and to facilitate or in some cases fully automate other verification processes. For example, if bugs were found where none were expected using random testing, the user could respond by updating the verification plan to include explicit functional coverage that targets the areas where the bugs were found.

The Automated Metric-Driven Process Model

Let's take a look at the process model in an automated metric-driven process. We'll discuss each phase individually outlining how each phase is related to the others and how each phase is enhanced by the new methodologies discussed in this book and by the application of MPA tools.

2 Automated Metric-Driven Processes

Planning

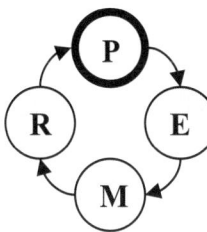

During the planning phase we will determine what needs to be verified and what metrics will be measured. There are many stakeholders in the verification process. Each stakeholder tends to have different concerns about the device. They each have their own perspective on verification. Some of these perspectives are shown in Figure 2.2.

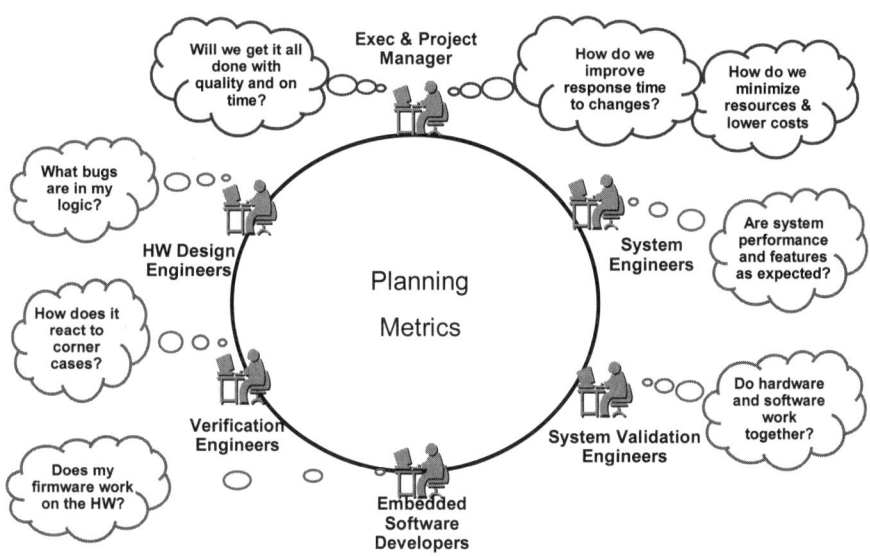

Figure 2.2 Stakeholder Perspectives

Verification concerns that are raised late in the project are one of the main causes of schedule slips. To avoid this, we involve *all* stakeholders in the planning phase. The planning technique used in the MPA methodology may be a bit different than what you are used to. It is a collaborative brainstorming effort. All stakeholders in the project participate in the planning session. During the session, the device is discussed on a feature basis. The designer presents his section of the device based on what it does. Each participant is encouraged to contribute their concerns about a given feature to the discussion. Along with each concern, the participant works with the group to identify a metric that will guarantee the concern was addressed.

Each of these metrics must be *objectively and automatically measurable*. By using objective metrics, we remove human subjectivity from the equation. We know the exact status of our processes based on the metrics we have defined. By using only metrics that can be measured automatically, we ensure that we will always have real time status. Tracking process metrics is no longer an "extra" task that may get lost in the shifting priorities of a hurried project.

Let's illustrate capturing concerns with a few examples. A design engineer is concerned that his DMA engine be exercised in such a manner that the input and output FIFOs are full simultaneously. An assertion that checks for this condition will provide a metric that addresses the concern (assertion coverage).

A verification engineer is concerned that every feature is exercised in every possible configuration mode. This concern can be addressed using functional coverage as a metric.

A firmware engineer is concerned that the DMA engine can move the appropriate OS code from the ROM to the instruction memory. This concern can be addressed using functional coverage as a metric as well. *Each of these metrics can be captured automatically from our verification engines.*

The output of the verification planning session is an executable verification plan. This plan will be used as the basis for determining what tasks should be executed as the project proceeds.

An example verification plan is shown on the next page. The concerns of the design, verification and software engineers are captured for a DMA engine in our device. The top half of the page shows the plan as it is written during the verification planning interview. The bottom half of the page shows the plan as it appears after it has been read into an MPA tool and the coverage metrics have been collected from several runs of our verification engines (Figures 2.3 and 2.4).

Features

1.1 DMA Engine

The DMA engine moves blocks of data between the various memories of the device and the external memory. The engine is configured via address mapped configuration registers.

Design
cover: /sys/rtlcodecover/dmamod/*

Verification
cover: /sys/verif/dma/regreadwrite/*

Software
cover: /sys/verif/dma/scenario/instructionmove/*

Figure 2.3 Verification Plan Editing View

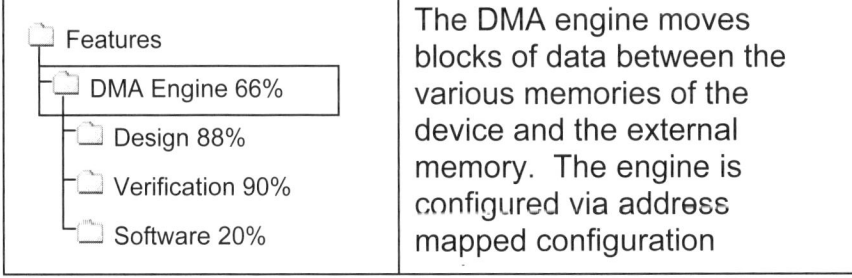

Figure 2.4 Verification Plan Executable View

As mentioned above, the planning sessions are collaborative. The value offered by the resulting plan is directly proportional to the number of stakeholders that actually attend the session. With the hectic pace of chip design projects, it is often difficult to arrange for all stakeholders to attend these sessions. It should be stressed that these planning sessions are *extremely* valuable. They enable all the other techniques that will be described in the MPA methodology. Because of the collaborative nature of these sessions, several device bugs have been found while planning, without writing a single line of verification code.

Management Planning
As the various stakeholders are outlining their technical concerns, management contributes by defining priorities for completion and schedule milestones. These priorities and milestones are captured in the plan as well. Priorities can be incorporated into the verification plan as weights on metrics corresponding to the key-prioritized features of the device. Most MPA tools allow milestones to be defined as well so that metric status data can be displayed along with defined milestones to make tracking the completion of the project more convenient.

For more information on defining weights and milestones see the chapter on planning in Part II of the book. It is very important to have appropriate reporting mechanisms organized and functioning *before* the project begins. Two of the key aspects of these reports should be the priority of different objectives and the milestones that are defined for their completion.

Visibility of the Plan
One of the key requirements for metrics to actually be useful is visibility. That means visibility to everyone. In order for projects to come in on time, we need to make it impossible for anybody to "massage the status" either intentionally or not. Objective metrics go a long way in this direction.

All available metrics should be made visible to *all* the project's stakeholders. By making these metrics available, we enable each contributor to creatively solve problems as they arise because they are aware of them. How many postmortems have you been to where an engineer said "Well, if we'd known what was happening, we could

2 Automated Metric-Driven Processes

have executed the following process to solve the problem?" Wouldn't it be nice if you never had to hear that statement again?

By automatically collecting objective metrics and allowing each of the users to personally interpret them we avoid two classic problems. First, automatically collected metrics do not create a resource drag on the project. No more walking from cubicle to cubicle to collect the daily status. No more interminable status meetings. We let computers do what computers are good at. As they run our simulation and emulation jobs, they automatically collect the metrics that we define as important. Second, we remove error-prone humans from the reporting process. The metrics collected are exactly and only the metrics created by our verification tools.

As discussed we'll use automatically measured metrics to gauge the completeness of all our processes. In the execution phase, we'll capture those metrics from our verification engines. For an in-depth explanation of the planning phase see Part II.

Execution

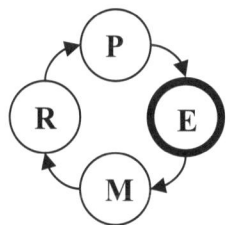

During the planning session, we captured every concern and corresponding metric in an executable verification plan. In the execution phase, we'll execute on those plans. There are two types of execution: implementation execution and verification engine execution. Implementation execution refers to the efforts made by engineers on the project team to implement environments that will run on the verification engines. Engine execution refers to the actual runs of the verification environments produced by the engineers.

Using a metric-driven approach allows us to improve the predictability, productivity, and quality of both implementation and engine execution. There are two opportunities that are presented by using metric-driven processes. First, using automatically capture metrics, we can get a better perspective on how our processes might be improved. Knowledge is power. Second, using automatically captured metrics, it is possible to fully automate some processes and remove the human element.

The execution phase at first glance seems simple, and it should. During this phase, we execute on our plans. During the execute and measure phases, our MPA tools will annotate the measured metrics specified in the verification plan to the metrics defined in the plan. Our verification plan will always have the latest status of all defined metrics embedded in the document. The flow for collecting metrics is shown in Figure 2.5.

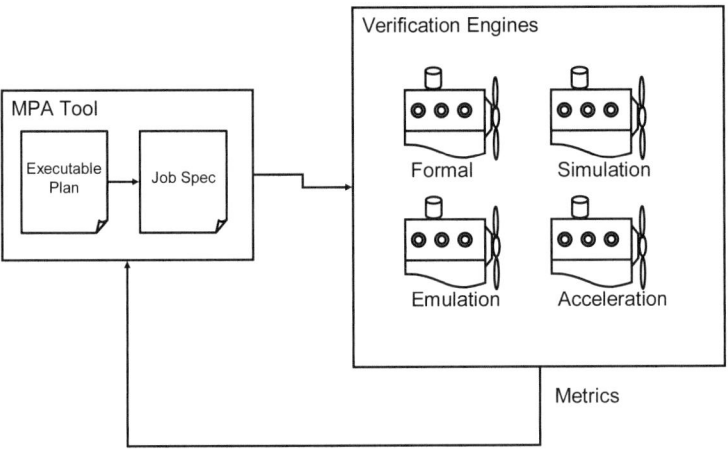

Figure 2.5 Automating the Execution Process

Using the MPA tool, we start our verification engines. The MPA tool automatically tracks the metrics specified in the plan by extracting them from the outputs of the engines. These extracted metrics are then annotated back into the plan.

It is possible to detect problems earlier and better utilize resources because metrics can now be automatically captured and analyzed. With some advanced planning, teams can begin to solve a number of problems by collecting data that illuminates both the causes and solutions of those problems.

2 Automated Metric-Driven Processes

Using metric-driven concepts some time consuming, tedious processes can even be fully automated. Generation of detailed debug information is a good example of this.

Simulation time is valuable and should be used efficiently. After implementing or changing a design, the engineering team runs large sets of simulation testcases, (called regressions), to ensure the design has been implemented correctly. Once the current implementation is deemed acceptable by passing these testcases, more implementation can begin. Because of the iterative nature of this implementation process, it is desirable that these test suites execute as quickly as possible. To increase execution speed, most of the debug features of the simulator are turned off. Typically the only failure or debug information available in this mode is a brief message describing the failure and the time that it occurred.

However, to completely debug an issue, an engineer needs much more information, such as waveforms that illustrate the signal levels of interest around the time the issue was detected. To gather this information, an engineer will manually sort through the failing testcases determining which testcases produce unique failures in the shortest amount of time and then run these simulations again with the waveform generation feature of the simulator turned on. This process is shown in Figure 2.6 with the human intervention points clearly marked.

By using our MPA tools to automatically analyze our captured metrics, we can completely automate this process. The technique is simple. The MPA tool captures the various unique failure types and then determines which testcases produced the failures in the shortest amount of time. Then, because the MPA tool has access to all the information required to start a given testcase, it can restart the pertinent simulations with debug features such as waveforms turned on. As a result, our engineers no longer spend hours analyzing failures and then waiting for the simulations to rerun.

24   Metric-Driven Design Verification

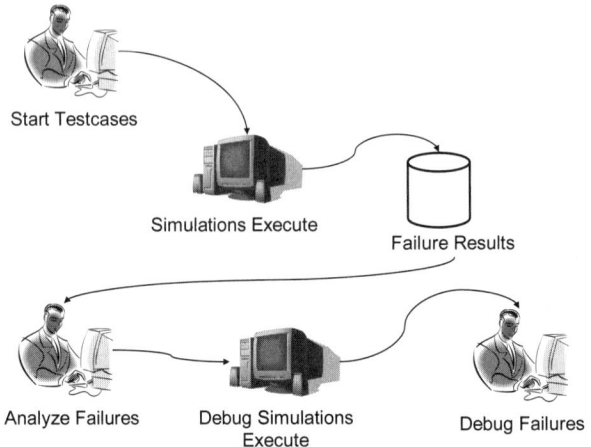

Figure 2.6 Manual Debug Process

They simply start the simulations once and then analyze the failure data as soon as the simulations complete as shown in Figure 2.7.

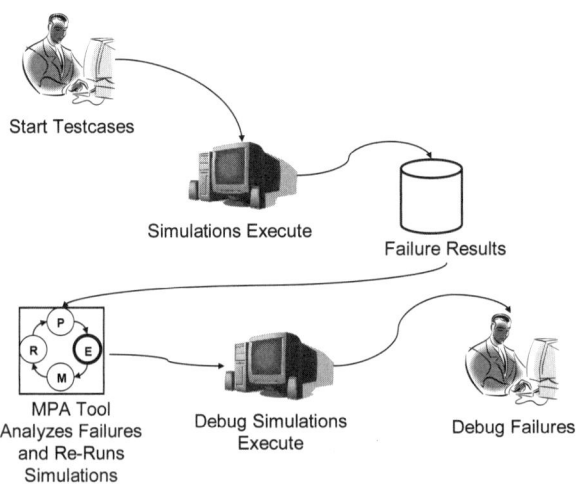

Figure 2.7 Automated Debug Process

Measurement

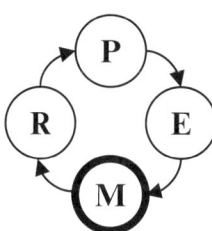

The measurement and analysis phase of the process is one of the most automated phases, and provides the bulk of the power offered by the MPA methodology. During the measurement phase, the MPA tool automatically captures and stores all the metrics that the project team has declared during the planning stage. MPA tools such as Cadence's Incisive Enterprise Manager ship with built-in metric-capture mechanisms for popular verification tools such as simulators and emulators. These capture modules automatically scan the output of simulator tools and extract common metrics such as failure messages, the amount of CPU time consumed by the simulator and the amount of real time consumed by simulator execution. By building easy to implement metric capture plug-ins, engineering teams can capture any other objective metrics from the outputs of their verification tools.

Using these metrics, MPA tools can automate standard analysis tasks as well. Using the example from the execution section earlier, from our simulation runs, the MPA tool captures:

- Failure type
- Failure time
- Testcase name

These metrics might be stored in a table (Table 2.1).

Table 2.1 Simulation Failure Metrics

Testcase Name	Failure Time	Failure Type
Dmaengine1	1000	FIFO pointer assertion
Dmaengine2	200	FIFO pointer assertion
Dmaengine3	1025	FIFO pointer assertion
Dmaengine4	257	Bad read/write pair

Using these metrics, the MPA tool can first group on the failure type and then sort on the failure time to determine the shortest set of testcases that can reproduce all the failures with debug information turned on. The results of this analysis are shown in Table 2.2.

Table 2.2 Failure Metric Analysis Results

Testcase Name	Failure Time	Failure Type
Dmaengine2	200	FIFO pointer assertion
Dmaengine4	257	Bad read/write pair

Engineers can define and store automated analysis tasks such as the one above. The MPA tool can then automatically perform these tasks where appropriate.

Response

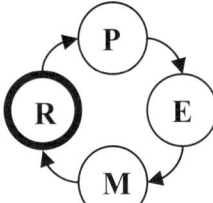

During the response phase, our human resources re-enter the picture to do what they are best at: develop innovative solutions to improve the status of the project as revealed by our automated metrics capture and analysis.

Using our MPA tools for automated analysis, we can get project status such as that shown in Figure 2.8.

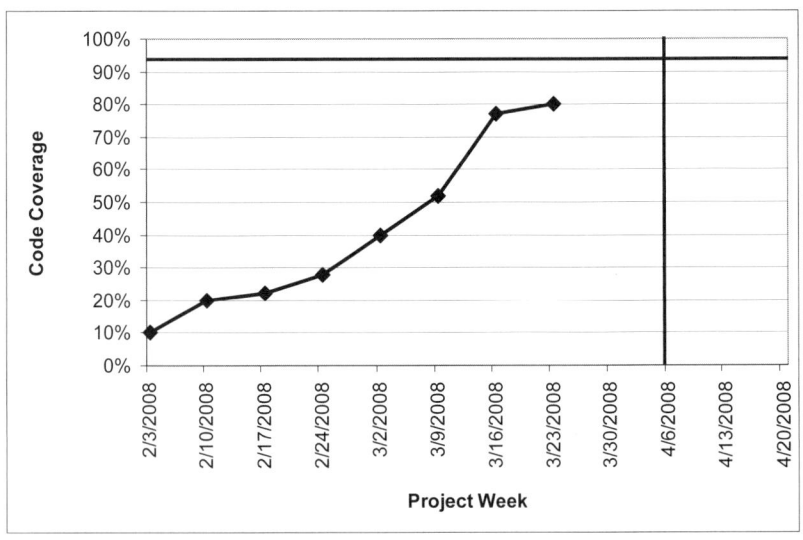

Figure 2.8 Project Wide Code Coverage vs. Project Week

With this information, our management team might judge that everything is on track. However, the data in Figure 2.9 tells a different story.

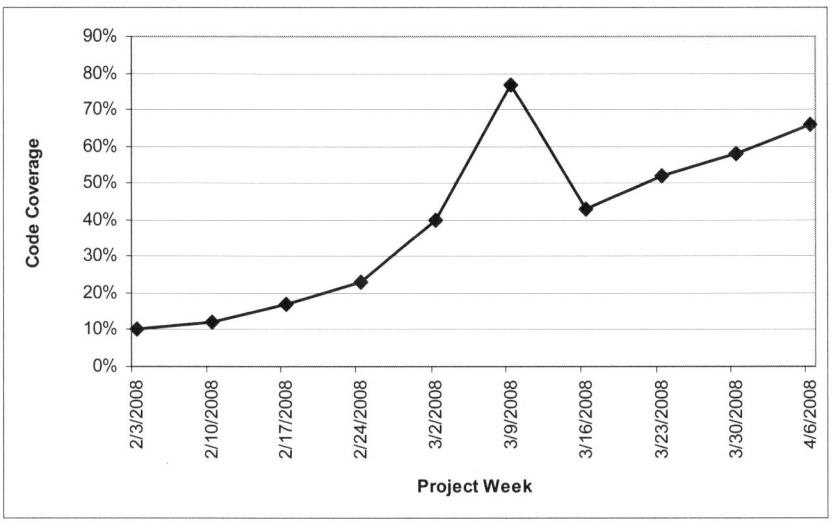

Figure 2.9 DMA Code Coverage vs. Project Week

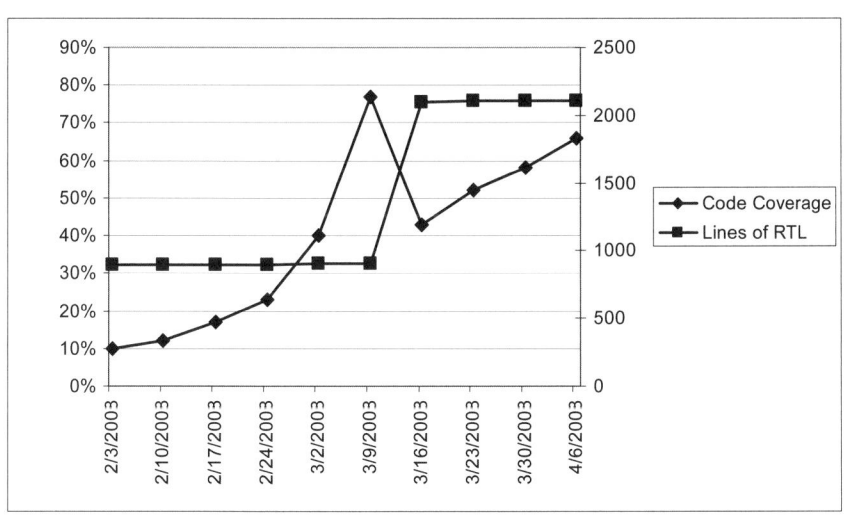

Figure 2.10 Code Coverage and Number of Lines of Code

We can see that there was a large reduction in code coverage for the DMA block on 3/16. By itself, this should be a red flag that indicates action needs to be taken with respect to the DMA block to avoid placing the schedule at risk. Using other automatically capture metrics, we can gain even more insight into the project status.

In Figure 2.10, we can see that the reduction in the total code coverage was caused by a large increase in the number of lines of code implemented. This may have been caused by a new feature being added to the design. In such cases, management can now make judgments regarding the cost of adding the feature based on real data rather than impressions or opinions. While decision should *always* be based on real data, it is much easier to follow this axiom when the real data is readily available.

Project Management Using Metric-Driven Data

Hundreds of books have been written on project management. Everyone has their favorite methodology. Almost all these methodologies have one thing in common: the raw material for making decisions is objective data collected from the execution of the project itself. This data is exactly what metric-driven methodologies provide. Rather than trying to expound on our favorite project management methodology here, we'll respect your choices. This book tells readers how to determine what metrics to measure and how to objectively and automatically gather those metrics to enable your project management methodology of choice.

As mentioned earlier, the paramount task before the readers is to make sure that their metrics are properly utilized. As part of the planning process, the management team should identify all status reports that they will need to properly evaluate and manage the project. These reporting mechanisms should be setup and tested *before* the project begins. This small amount of effort early in the project will enable all the gains discussed throughout the remainder of the book.

What Are Metrics For?

Knowledge is power! Our metrics will increase the power of the project team in several ways. First, metrics give the team the chance to react to dynamic requirements changes. As resources and requirements change, subsequent changes in measured metrics allow us to detect and react to these changes. Second, as illustrated above, metrics can be used to completely automate some processes.

Tactical and Strategic Metrics

In this book, metrics will be placed in two broad categories. The first set of metrics is tactical metrics. These are metrics that give the team data about where the project stands at present. Tactical metrics, as their name implies, are used to make tactical decisions during the course of the project. Examples of tactical metrics are the number of testcases that failed in the most recent regression, the percentage of coverage completeness on the serial block, and the number of assertions that activated successfully for each module in the chip.

The second set of metrics is strategic metrics. These are frequently referred to as historical metrics. Historical metrics may start out as tactical metrics, but in this context, they are tracked throughout a project and then used to make strategic decisions at a later stage of the project, or in a follow-on project that reuses aspects of a previous project. Historical metrics include the number of issues found in a module during development, the rate at which coverage closure was reached on a project, and the frequency that a module was revised over the course of a project.

Historical metrics can also be used to shape training and career development plans for members of the engineering team. They can be used in much the same way that professional athletic coaches use game statistics to determine where to focus the next weeks' practice sessions. We'll talk more about this in the verification management chapters.

Summary

In this chapter we have illustrated the evolution of and justification for a metric-driven verification process. This process can be divided into four steps as shown in Figure 2.11.

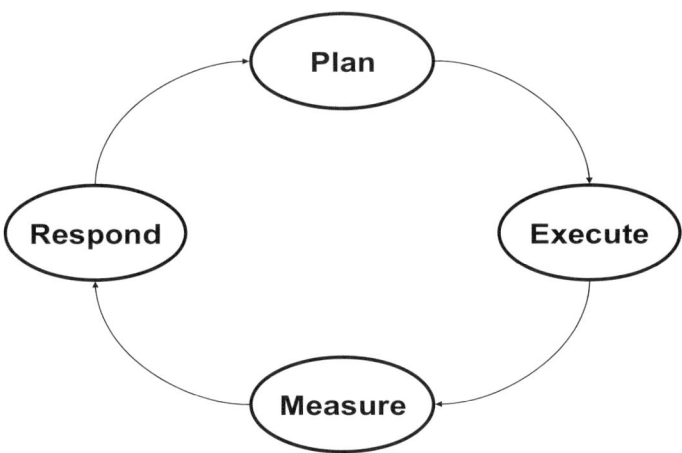

Figure 2.11 The Verification Process

We will use this diagram throughout the book to help illustrate what portion of the process is being described. We looked at some of the key points of each of these four phases, and showed how the MPA methodology offers improvements in each phase.

Chapter 3
Roles in a Verification Project

Introduction

In this chapter we will describe the various roles in a verification project. How each role fits into a metric-driven verification process will be illustrated. We'll discuss what metrics each contributor produces and consumes and how they can best analyze the metrics that they consume.

Verification teams have a number of diverse roles requiring somewhat disparate skill sets. These roles are described below. One person may serve in more than one role.

You may be surprised to see some of the roles that are included below. However, each of them contributes something vital to the efficient completion of the project. Depending on the company culture and various aspects of the design project, all of the roles may not exist in every project. In many projects, several roles are rolled (pun intended) into a single individual that prefers a given title. This explains the preponderance of companies that "don't have a verification team." They of course have one. The designers would just appreciate it if you didn't call them so.

The Executive

Yes, executives actually do have a role in design and verification engineering! The size and nature of this role will vary based on the size of the company and the style of the executive. For small startups, the executive may be very involved in their companies' only revenue generating project. Larger, more-established companies may have hundreds of design projects forcing the executive into a necessarily more diluted role on a project by project basis. In any event, the executive is often the largest individual stakeholder in a given design project.

The important thing to note for our work is that even the executive should be able to view an appropriate set of aggregate metrics from every design project.

At the executive level, the metric of most interest and importance is the completion status of the project. There are three main components of completion status. They are:

- Implementation completion
- Verification coverage
- Activity indicators

Implementation Completion
The implementation completion metric is a measure of how much of the planned work is implemented. For example, if one hundred testcases have been defined for a given part of the design and 80 of them have been completed, then the implementation completion metric for that task is 80%. MPA tools can provide roll-ups of metrics so that the executive can first view results for the complete project and then hierarchically investigate results for each portion of the project individually (Figure 3.1).

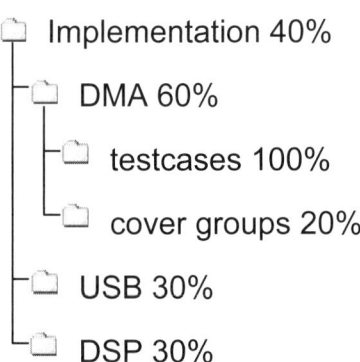

Figure 3.1 Implementation Completion

Verification Coverage

Implementation completion is a measure of how much of the specified work has been completed. Verification coverage is a measure of how well that completed work is verifying the design vs. the specified verification concerns. In verification planning design and verification engineers, firmware and application engineers, and other interested stakeholders meet and determine what features of the chip must be tested and how. The verification environment is the vehicle that will be used to verify the chip. Verification coverage is a measure of how many of the required cases have been verified (Figure 3.2).

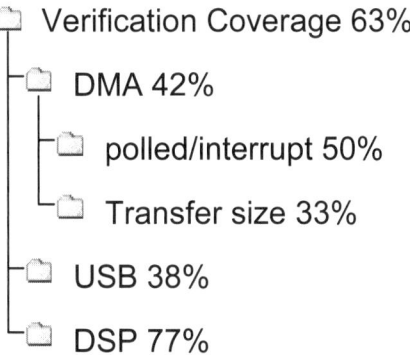

Figure 3.2 Verification Coverage

Activity Indicators

An activity indicator tells the executive where the team is spending their time. Are they implementing design or verification environment code? Are they writing testcases? Are they debugging failures? Metrics that are produced by revision and issue-tracking systems and processed by an MPA tool can provide this information. The data can be grouped by engineer or team to determine what activity is consuming most of the team's time, or it can be grouped by design block to determine what stage of development a given block is in.

Marketing

In-bound marketing is primarily focused on benefits to customers and requirements of customers in different market segments. Marketing metrics are largely the same metrics that interest the executive. They may be organized differently however. Except in companies where marketing takes on the program management tasks as well, marketers may be less interested in views of project data from an engineer or engineering team perspective. Marketing will be more interested in looking at metrics with respect to the important customer-driven features of the device. The exact same data can be displayed in a different perspective so that marketers can track progress against their most important features.

Design Manager

The design manager leads the design team in their implementation of the device. Depending on the size of the project team, the design and verification managers may be one and the same person.

Metrics Used
Design-specific metrics are of interest to the design manager. Among these are:

- Number of assertions written per design module
- Trend of revisions made per design module
- Number of comments inserted per design module
- Issue detection rate
- Code coverage closure

Verification Manager

The verification manager serves as the verification project coordinator and facilitator. The verification manager need not be the technical lead as well. However depending on the size of the project and the size of the team, the verification manager often serves as the technical lead.

As the key custodian of verification data, the verification manager should be familiar with all the other stakeholders of the chip project and their perspectives. It is necessary for the verification manager to be familiar with the design, verification, and production processes. The verification manager should also be aware of how each stakeholder is making use of the data and metrics created by their team. This includes stakeholders at all levels of the hierarchy. A familiarity with how executives such as the CEO and Business Unit Director are making use of the metrics produced by the verification team is just as important as knowledge of what the verification team members doing with them.

The verification manager will make use of both tactical and strategic metrics as the project proceeds from the planning to the implementation to the execution and closure stages. They will look at a more abstracted version of the tactical metrics that tend toward the strategic. Some of the metrics used by the verification manager are:

- Utilization of verification tool resources
- Closure rate for coverage metrics
- Number of test scenarios created and running
- Issue detection rate
- Module revision rate

These metrics will be used to make short-term strategic decisions about how to apply verification resources to best meet the schedule and quality requirements for the project.

Verification Architect/Methodologist

The verification architect lays out the abstract descriptions of the various verification environments and defines the verification methodology for the project. The recommendations of this contributor will have a profound impact throughout the project. They are of key importance during the planning and execution stages.

While the architect may not do much of the actual verification coding, they are one of the key contributors to the verification project. Whether or not the device is completed on time with high quality depends on good architecture and methodology as much as any other aspect of the project. This individual should be very experienced, and have a command of most if not all aspects of verification technology that are to be used in the project.

During the planning stage, the architect influences what activities will be executed as the project proceeds. There are several types of metrics that the methodologist will use to perform their job. These are:

- Historical reuse metrics
- Historical engineering resource metrics

The information that the architect/methodologist will call on include:

- Available IP and VIP
- Available engineering resources
- Available verification technology
- Historical metrics attached to VIP/IP
- Historical metrics attached to similar projects

The verification methodologist should work closely with the verification team to determine what verification technology should be used taking several things into account. Early in the project, historical metric data may be of the most use to the verification methodologist. By using this data, they can determine what technologies, IP, and methodologies can be best deployed to complete the project within the desired schedule.

As the project proceeds, the architect's role may not be as intense as during the planning stages, but they still serve a very valuable purpose. They can help steer the verification team members to efficient implementations through mentoring activities such as code reviews, and consulting.

Design/System Architect

The design or system architect is interested in how the architecture specified for the device is performing as real hardware is integrated into the device. Checking that real hardware performance matches architecturally planned performance is of particular interest to the design architect. In addition to simply testing performance criteria, an architect may also identify scenarios that place the system under stressful conditions. They will be interested in seeing that these scenarios have been exercised. They can derive the coverage status of these scenarios from functional coverage.

Some of the metrics used by architects are:

- Performance testing results
- Functional coverage

Verification Engineer

The verification engineer works on the day to day implementation and iteration of verification environments. This is one of the most challenging roles of the entire project (guess what the authors used to do for a living!). In this role, it is not enough to be technically literate with the given verification technologies. More than any other role, the verification engineer is required to have a complete understanding of the operation of the device and how it is to be used in systems.

The verification engineer is forced to make the switch from code implementer to problem solver mid-stream in the project. The first portion of the project will entail implementing in the most efficient manner possible every kind of test that will exercise the device. As soon as this activity is complete, the verification engineer then switches to intensive problem solving to determine what if anything is wrong with the current execution of the design under the influence of these test scenarios. In a typical project, this abrupt mindset change will be required not once, but repeatedly as the project iterates from block to unit to chip- and system-level activities.

The verification engineer makes use of most metrics in their raw form. Some of the metrics used by verification engineers are:

- Functional coverage
- Assertion coverage
- Hardware code coverage
- Software code coverage

- Revision control data
- Historical regression data
- Tactical regression failure data
- Historical VIP metrics
- VIP documentation
- IP documentation

These metrics will be used on a daily basis and efficient collection of and access to them is essential to the verification engineer's success.

Design Engineer

The design engineer has what might be considered the most responsible role of the entire project. They deliver the code that will eventually become the device that is shipped to the customer in return for revenue.

Design engineers consume many of the same metrics that verification engineers do, but not always for the same reason. For example, a verification engineer uses functional coverage information to determine where to steer random constraints so that more of the device can be tested. A design engineer might use the same information to determine how complete the verification effort is, or to determine the set of testcases that best exercise a given portion of the design.

The metrics used by design engineers are:

- Functional coverage
- Assertion coverage
- Hardware code coverage
- Revision control data
- Historical regression data
- Tactical regression failure data
- Historical VIP metrics
- VIP documentation
- IP documentation

Regressions Coordinator

A regression is a set of verification environment executions performed to test the device. A verification environment execution can be anything from running a formal verification tool, to running a dynamic simulation, to running a hardware-based emulation.

The regressions coordinator is responsible for managing the regression process. Their duties include managing software and hardware licenses, and automating the execution of regressions and the analysis of the data returned from regressions. Leading edge technology can help automate several of these tasks. Depending on the size of the project and the diversity and quantity of verification resource available, this role can be a part-time responsibility.

Regression coordinators use a different set of metrics. They are interested more in the overall flow of the project and less in project specifics. The metrics they use are:

- CPU utilization
- Emulator/Accelerator utilization
- Software license utilization
- Coverage gained per regressions
- Issues found per regression

The first three metrics tell the regression manager whether the resources at their disposal are being utilized efficiently. The last two give an indicator as to whether or not regression activities are effective.

Debug Coordinator

The debug coordinator is responsible for analyzing failure information returned from regressions and then assigning those failures to various design and verification engineers for debug. This position is seldom a job in its own right. Frequently this duty is shared among the verification team members.

The debug coordinator is primarily interested in failure metrics. Failure metrics include:

- Failure description
- The verification tool that generated the failure
- The time required to reach the failure
- The portion of the design in which the failure took place
- Revision control data

Summary

In this chapter we briefly outlined the responsibilities of each verification stakeholder. We also looked at what metrics each stakeholder uses and how they put these metrics to use.

In Chapter 4 we'll present an overview of the activities that take place in a typical verification project.

Chapter 4
Overview of a Verification Project

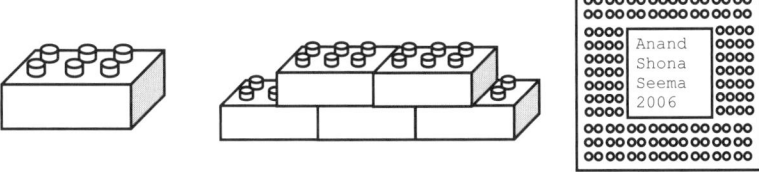

Introduction

This chapter outlines the flow of a typical verification project (Figure 4.1). As a chip design moves through several developmental stages as it evolves from a concept to a finished product. These stages are:

- Marketing definition
- Architectural exploration and specification
- Block-level implementation and verification
- Integration verification
- Chip-level verification
- System-level verification

We'll explore what tasks are performed at each stage, what metrics provide visibility into the status of those tasks and who performs them. As we describe each task, we'll also mention various tools and techniques that are available to execute the task. This chapter is intended to be read in conjunction with the following chapter that provides a deeper explanation of the various verification technologies available.

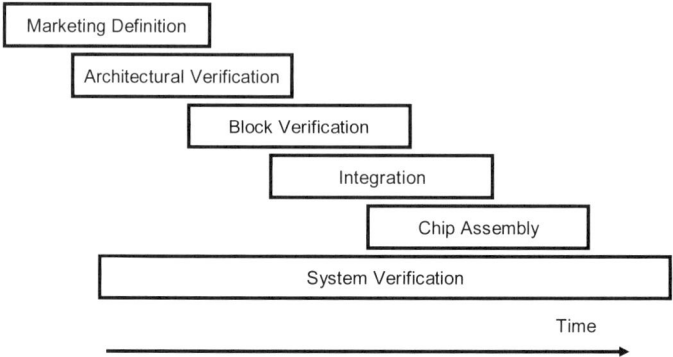

Figure 4.1 Verification Stages

Marketing Definition
In this stage, the business unit decides what features will sell the device. Decisions are made based on a combination of customer desires and engineering capabilities. The outputs of this stage are the device features and the intended functionality of those features. This is where product intent if first defined.

Architectural Verification
At this stage system architects perform simulation studies to determine what the optimal configuration of the system to be designed is. This is often the first translation from the intent of what the device will do to how it will be implemented albeit in very wide brush strokes. The key output of architectural modeling should be a set of architecture decisions embodied in a functional specification. Design and verification engineers will use this specification to implement the device and its verification environments.

Architectural verification can be done at all levels of device integration. Specific studies may be made at the block level for crucial portions of the device. Other studies may be made at the chip level to determine, for example, if the communications architecture of the integrated chip will offer sufficient bandwidth. As with other types of verification, the engineers should consider how to reuse verification environments across integration levels.

Some architectural studies such as communications bandwidth modeling can be done with simple tools such as spreadsheets. For more detailed studies, some level of simulation is used. Transaction-level modeling (TLM) is used to simulate abstract models of the device very quickly in software. TLM models are most often written in SystemC, however they can be written in any behavioral programming language.

When using TLM simulations, the model can be verified using traditional simulation testbenches and assertion-based techniques. The verification at this level is necessarily coarse. Only the details that influence architectural decisions such as bandwidth capacity or algorithm output are checked.

The metrics produced by this activity are:

- Functional coverage
- Assertion coverage and/or check coverage

Functional coverage is used to gauge the completion of the architectural verification with respect to the scenarios that the project team identified during planning. Assertion and checking coverage is used to ensure that all checks that were defined are actually implemented and have been exercised.

Careful planning and implementation of architectural verification IP leads to many reuse opportunities. The reusable IP includes:

- Assertions
- Transaction-level architectural models
- Transaction-level testcases and testbenches

Assertions used at the architectural level can be remapped to be used for hardware verification at the block, chip, and system levels as well. Reusing these assertions, we can verify that architectural assumptions are valid as the device moves closer and closer to the final production stage.

The models of the device used to perform architectural studies are frequently accurate enough to be used to check device behavior in the later stages of verification. For example, suppose the architecture team is determining how to split the various steps of a graphic decoder algorithm between hardware blocks. They chose a partitioning as shown in Figure 4.2.

Figure 4.2 Algorithm Partitioning

Each step of the algorithm will be done by an individual design block and the design blocks will share a common communications bus. The architects must determine if the bus has sufficient bandwidth to support the traffic that flows from block to block without interruptions. The block-level architecture of the device is shown in Figure 4.3.

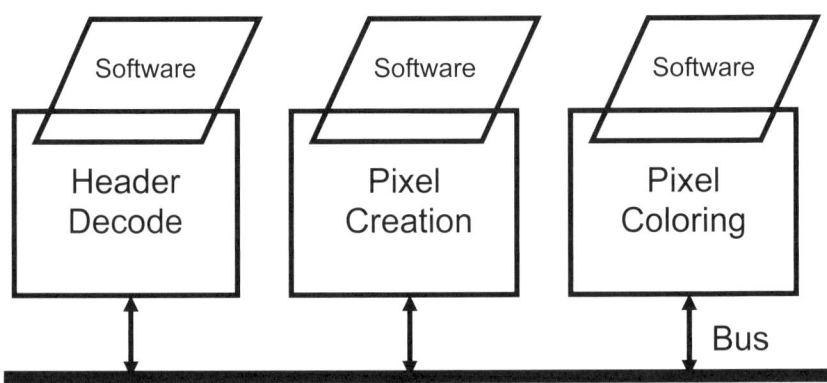

Figure 4.3 Block-Level Architecture

4 Overview of a Verification Project

To generate the data that will be transferred from block to block, they will actually use a software model of the algorithm. The software model will be partitioned in the same manner and embedded in the transaction-level simulation models for each block as shown in Figure 4.3. As the simulation executes, the software model will create data and this data will be passed on the bus as transactions by the hardware models. The architectural engineers can then check the bandwidth usage vs. their assumptions.

When the actual design is verified, it will be necessary to check that the design is producing the correct data for each step of the algorithm. Rather than writing a new checking model, the verification engineer can simply reuse the software models that were created for the TLM simulations by embedding them in their verification models. In a similar manner, streams of transactions that are developed for architectural simulations can be used to stimulate actual hardware devices as well.

The tools most frequently used in architectural verification are software simulators that use TLM level models. As actual hardware blocks of the design (register transfer language, or RTL models) are completed, TLM blocks can be simulated in the same environment with them. Although emulators and accelerators are rarely used for architectural modeling, TLM blocks can communicate with the hardware blocks that are modeled by these technologies as well.

Block Verification

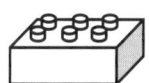
Once the system architects produce a basic block diagram of the device and a functional specification, design engineers begin to implement the hardware that will become the device itself. Typically this hardware is created in a modeling style called "register transfer language" (RTL). RTL can be created in a number of modeling languages, including Verilog, VHDL, and System Verilog. The RTL models created using these languages can be simulated on software simulators, or accelerators and emulators. Verification of these models may be done using:

- Formal assertion-based techniques
- Dynamic assertion-based techniques

- Dynamic software simulation testbenches
- Emulator or accelerator testbenches

While some verification is performed by design engineers themselves, the bulk of the verification at this and subsequent stages is performed by verification engineers.

The key metrics used to track verification closure at this level are:

- Functional coverage
- Code coverage
- Dynamic assertion and/or checking coverage
- Formal assertion coverage

The key output of this type of verification should be a qualified block design that is ready to be integrated with other blocks to create sub-systems, or the complete chip itself.

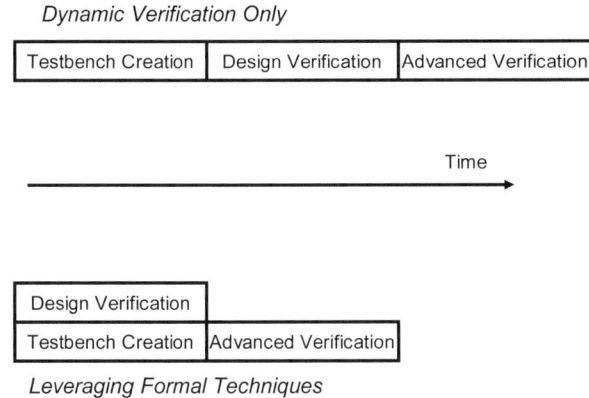

Figure 4.4 Schedule Savings Using Formal Techniques

Using formal verification tools (Figure 4.4), the designer can begin to verify their block before any verification infrastructure is available from the verification engineering team. This is a huge improvement. Previously, there was often a delay in the schedule as the designer waited for a verification environment that could exercise their device. With formal techniques, the designer can begin verification using assertions (concise declarations of behavioral rules), as soon as they begin coding their module.

Formal verification techniques are best aligned to blocks of control logic as opposed to blocks of logic that perform data transformations such as multipliers. Because they perform state space explorations, formal verification tools are best suited to smaller blocks of logic.

As formal tools run out of steam, verification engineers use dynamic simulation and testbenches to further verify the device. Simulation testbenches simulate the actual operation of the device under test and its surrounding environment. Using simulation, engineers can model input transactions that stimulate the device as it will be exercised in a real-world system. They can also monitor output transactions to check for correct device behavior. Engineers can build streams of transactions that model real-world scenarios. The simulator also allows the engineer to monitor or drive any signal within the device.

Emulators and accelerators can be used to simulate block-level models. However, this is not done frequently.

Integration Verification

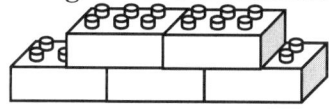

Integration-level verification is performed to check that subsystems within the chip work well together. As blocks are qualified as viable by block-level verification, they are introduced into integration-level verification. The key difference between integration and chip-level verification is semantic. At the chip level of verification *every* block is included in the device model.

The completion metrics for integration-level verification are identical to those used for block-level verification. More attention is paid to functional coverage at this level because the scenarios of interest are more easily modeled and tracked using functional coverage.

The verification technologies used are also the same. Far less formal verification is done at this level as the design grows in size and begins to outstrip the capacity of formal verification engines. The assertions written for formal verification at the block level can be reused here though.

Dynamic simulation is used prevalently at the integration level. Depending on the size of the integration, simulation performance may begin to slow toward unacceptable levels. Integration is the first level that most projects begin to consider using acceleration or emulation technologies.

Testbenches written for integration verification can be reused for chip and system verification. If a simulation testbench is to be used in an accelerator or emulator, then special considerations need to be made to ensure that it works in both on the simulator and the accelerator or emulator.

Chip-Level Verification

As mentioned above, chip-level verification is a special case of integration verification with the entire design present. This verification activity provides us with the first indicator of quality for the overall chip. At this level, simulations run significantly more slowly, and more emphasis is placed on acceleration or emulation. Firmware routines may be tested on the device at this level of verification.

System-Level Verification

System-level verification tests the implemented device in an environment that approximates the target environment the production-level device will be used in. In system-level verification, we verify that the device will work with the actual firmware and software that will be delivered with it and that the device will work in the presence of typical transaction streams on its inputs.

The metrics used to track system-level verification are:

- Functional coverage of input and output transactions
- Assertion coverage
- Code coverage of firmware and application software

Less emphasis is placed on tracking functional coverage of scenarios within the device. These scenarios are much easier to verify and debug at the chip level of verification and below because of visibility and speed issues.

Assertions can be loaded into both accelerators and emulators. However, at this level they are typically not used as completion metrics, but rather as tools to help speed debug once an issue is found. These are the same assertions that were used at the block, integration and chip level, and therefore have a very fine level of detail. They are typically turned off until an issue is detected at the system level. They are then turned back on in an attempt to isolate smaller issues that may have led to the system-level failure. In this manner assertions can help engineers isolate the root cause of system-level issues much more quickly.

System-level verification can be performed at two different levels of abstraction. It is typically performed using the complete design in either simulation (very slow), or acceleration or emulation. System-level verification can also be performed using a TLM simulator. The same TLM model that was described in the architectural verification section sometimes offers high-enough performance to allow very early testing of real firmware and application software. The key to success with a TLM model for system verification is to model at a high-enough level of abstraction to not significantly impact the speed of the simulator.

Summary

In this chapter we have given a brief outline of the various verification activities. We have described each activity presenting the metrics that are used to track the activity, the tools used to produce those metrics and the participants who are responsible for completing the activities.

Chapter 5
Verification Technologies

Introduction

This chapter deals with the execution stage of the verification process outlined earlier and shown again below. In the execution stage we run our verification engines to verify the correctness of our device.

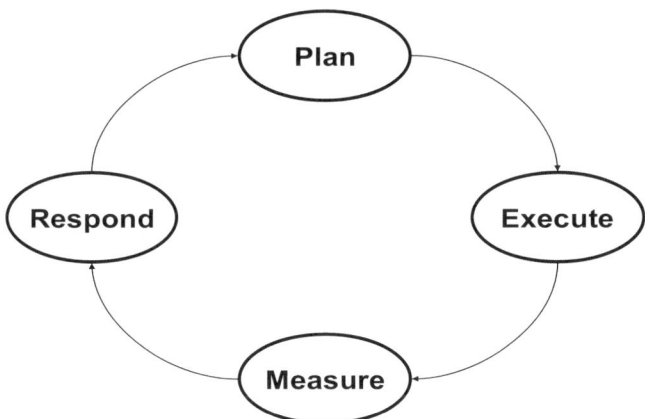

Figure 5.1 The Metric-Driven Process

The purpose of this chapter is to provide a high-level overview of each of the verification engines or technologies. This overview of each technology will include:

- Why to use the technology
- How the technology works

- How to plan using the technology and what metrics should be tracked
- Who utilizes the technology
- What process metrics should be analyzed
- How the technology can be leveraged

There are many different verification technologies or "engines" available. Each technology is suited to a given domain of verification that other technologies may not be. There is of course some overlap between these technologies.

Each of these technologies will be covered in more detail in Part III of the book.

Metric-Driven Process Automation Tools

This is the class of tools that enables all of the methodologies and concepts described in this book. These tools enable the "Plan, Execute, Measure, and Respond" metric-driven process framework shown in Figure 5.1.

Up until now, we've alluded to their capabilities. Here we'll dive into what they can do for you, how they work, how to utilize them, and the various use models of these tools.

What they are Used for
MPA tools enable the simple four-step process infrastructure that is described throughout this book: Plan, Execute, Measure, and Respond. They provide an automated framework that allows the user to capture the plan, control execution resources, measure the data or metrics created by those resources and then through either automated means, or human analysis and feedback to respond to the current metrics produced by iterating the plan.

5 Verification Technologies   53

Planning

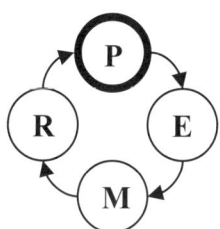

There are several types of planning that are currently in use. The purpose of all planning procedures is to document a course of action, and plan how the device will be verified. Two of the most used planning strategies are:

- Specification-based planning
- Feature-based planning

In a specification-based plan, the plan is created using the device specification as the raw material. The specification is reviewed and each point in the specification that is deemed important is recorded in the verification plan along with some metric to judge the completion of verification of that point. Specification-based plans may either be created by an individual and then reviewed by the team, or created by the team in a collaborative effort.

The difference between a specification-based plan and a feature-based plan is subtle, but important. Feature-based plans are always created in a collaborative way. All the stakeholders meet to create the plan. When we say all the stakeholders, we mean not only the verification and design engineers for a given block of the design for example, but also the firmware and application engineers, the system architects and even the design and verification engineers for other portions of the chip.

Take a look at Figure 5.2. It shows the various intent translations when moving from the original design intent to actual silicon. The original product intent is documented and then translated by one or more system architects into the architectural intent. This is then documented into a specification where it is translated by the hardware, software, and verification engineers of the project. By the time the actual design and verification environments are created, they are the products of at least three individual translations. By doing collaborative feature-based planning, you allow the various translators to resynch their resulting translations vs. the actual intent.

Either type of planning can be done as a one-shot process or an iterative process. In reality, planning will always be an iterative process. The circumstances of the project will change, people will hire on or leave, features will be added to or removed from the device, etc. The one-shot planning process attempts to account for all aspects of the project up front and then adapt to changes as necessary. It creates a "finished" document that will be ideally used throughout the project to define what actions are to be taken and how to track status.

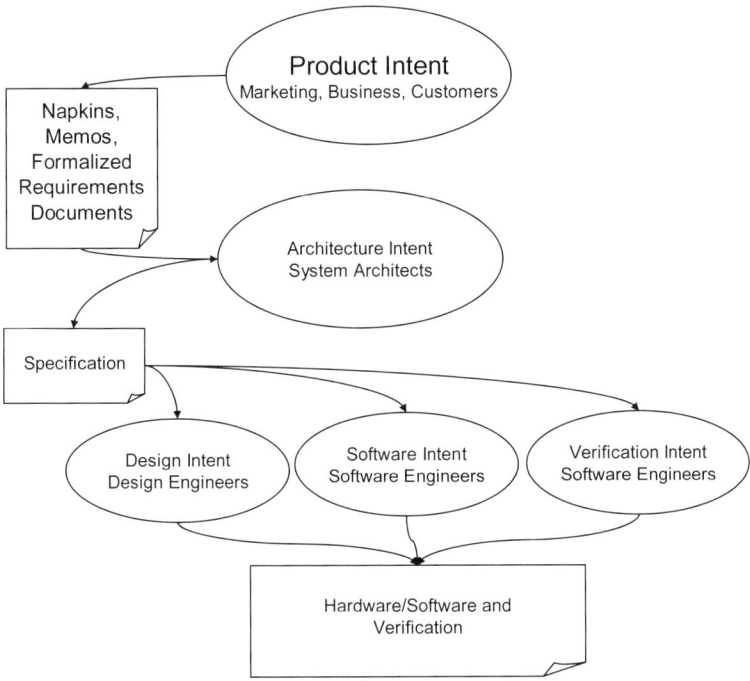

Figure 5.2 Product Intent Translations

An iterative plan takes these project changes as given from the start. The team creates a "good enough" plan to get everyone working and fully loaded. The team agrees that periodically (often every two weeks) they will meet and update the plan. This planning methodology places a lighter load on the team initially and tend to promote the survival of the plan. Because the team knows ahead of time that the plan will change and that the plan will be modified frequently, it tend to actually use the plan more often because it is more accurate.

5 Verification Technologies   55

More information on the planning process can be found in Part II of this book.

Regardless of the planning strategy used, the MPA tool allows you to capture the plan in an executable format. What's an executable format? This means the metrics that are to be measured to determine the status of the plan's execution will be automatically annotated back into the plan. This means we have closed the feedback loop on status tracking. Before MPA tools were available, the project team wrote a plan and defined how to track the status of that plan. Often the definition of success was implicit at best. In other words, the team didn't define a measurable metric for completion. They simply said that the plan was complete when all the testcases were finished. While this was simple, it was very subjective and very susceptible to human error.

With MPA tools, we define what metrics define completion of a certain aspect of the plan. The only qualifier to these metrics is that they must be automatically produced by one of our execution resources. Examples of metrics that indicate completion are:

- 100% code coverage of a design block.
- 100% functional coverage of all block-level covergroups.
- 100% coverage of all designer defined assertions.
- 100% of testcases defined for the block have been run and passed.

Usually a combination of all these metrics is used. The important aspect is that each of these metrics can be automatically measured. The term "executable plan" means that the measurements can be automatically gathered and attached to the appropriate portion of the plan so that you only have to view the plan to know the objective status of the project.

The excerpt (Figure 5.3) from the definition view of an executable plan serves as an example.

AHB Interface Block
    Code Coverage
```
Cover: /sys/ahbintf/codecover
       Covergoal:    100%
```
    Transaction Type Coverage
```
Cover: /sys/ahbintf/transtype_*
       Covergoal: 100%
```

Figure 5.3 Verification Plan Definition

The executable view would look something like Figure 5.4.

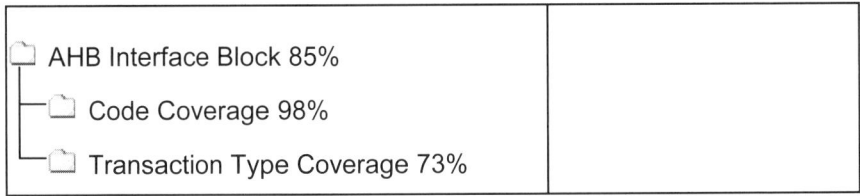

Figure 5.4 Executable View

The job of the MPA tool with respect to executable plans is to find the metrics specified in the definition view and display them appropriately in the execution view.

Execution

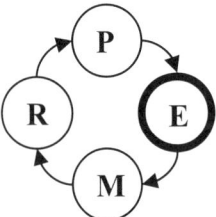

During planning, we define not only what is to be done and how to measure completion, but also how the tasks are to be executed. From an MPA point of view, this means we define what execution engines will be used to exercise our hardware design and verification environments. These engines might be simulators, emulators, linting tools, or any of the other various tools that are at the disposal of hardware/software/verification teams. The only requirement that MPA places on these tools is that they must be controllable via an automated mechanism (a scripting language for example), and that they produce metrics that can be automatically obtained by the MPA tool.

5 Verification Technologies   57

Depending on the MPA tool, different levels of refinement will exist. At a minimum, the tool allows the user to specify what execution engines are to be run and to also specify where the resulting output of the tool will be deposited so that the MPA tool can extract metric data.

The general execution flow of MPA tools is shown in Figure 5.5. The user supplies a set of job specifications that the tool interprets to control the various execution resources at the user's disposal. The tool then automatically parses failures and coverage metrics from the output of these tools and annotates those metrics back into the executable plan.

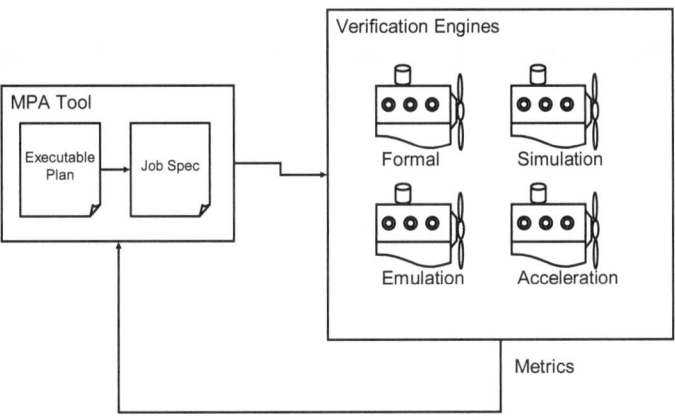

Figure 5.5 MPA Execution Flow

Respond

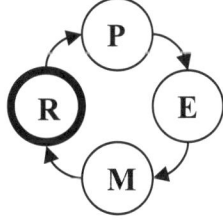

An MPA tool will provide two frameworks for response. The most common framework relies on human intervention and supplies a set of analysis tools. The simplest analysis tool is the annotated verification plan itself. By reviewing the annotated plan, the user can determine what tasks have and have not been completed. Based on this analysis, the user might respond by shifting resources within the project, or changing the scope of the project. A more refined MPA tool might allow the user to graphically explore what functional coverage was missing. By performing this analysis, the user can prepare reports that engineers can use to decide which testcases to write next.

The automated response infrastructure is less common, and a bit harder to use, but very powerful. Using this mechanism, a user can write small programs that read in the metrics harvested by the MPA tool, automatically analyze these metrics to create new job specification files, and then use these job specification files to automatically restart the execution engines to obtain better or more refined metrics. For example, an application can be written that first groups simulation failure types based on error messages. These grouped failures could then be sorted to find the simulation that has the earliest occurrence of each failure type in terms of simulation time. Finally, the application can create a job specification file that runs these simulations again with waveform creation turned on. This saves the verification engineers the time consuming step of resimulating failed simulations to produce useful debug information.

Now that we've looked at the methodology enabling engine, let's take a look at the other verification engines.

Modeling and Architectural Exploration
Why Perform Architectural Modeling?
Architectural modeling is in fact the first activity of design verification. It seeks to determine if the conceived architecture of the chip will provide the desired functionality and performance required by the product intent for the chip as established by the end customer.

Planning for Architectural Modeling

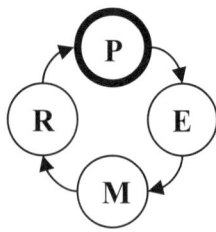

Verification planning consists first describing each feature of the device and then describing exactly *what* it does, and how to know *when* it did it.

Architectural modeling tend to be more concerned with performance trade-offs as opposed to

5 Verification Technologies

functionality. At this level of abstraction, it is assumed that the device will behave in a functionally correct manner. The key verification task here is to verify that the device has the appropriate resources organized in the appropriate manner to perform the feature correctly. Some examples of architectural "whats and whens" are:

- The device receives input data from the peripheral bus at an average bandwidth of no less then 64 KB per second. This average should be consistent over a time period no smaller than a tenth of a second (in other words, shipping no data for a second and then shipping 128 KB of data in the next second isn't good enough).
- The device caches intermediate data from the steps of the algorithm in no more than 64 KB of memory while flushing the memory in between separate executions of the algorithm.

Architectural modeling, like any other verification activity requires three dimensions:

- Determine what stimulus needs to be provided to the model to test architectural assumptions (stimulus).
- Determine how to measure that the stimulus and scenarios were in fact driven to the model (coverage).
- Determine how to check that the model functioned as intended (checking).

In short, architectural modeling is no different than any other verification activity. The level of abstraction is simply elevated to system-level concerns.

How It Works

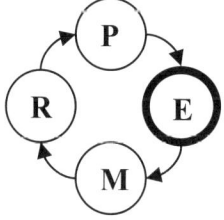

Architectural modeling helps architects to make decisions regarding the basic structure (architecture) of the chip. This is done by providing a very fast, abstracted simulation platform that allows the architect to model the behavior of the finished device at a high-functional level without the necessity of specifying all implementation details.

Architectural modeling has been rather well specified by the OSCI standard. There are several levels of abstraction that can be applied to the process. The higher the level of abstraction, the faster the underlying simulator will execute. The levels of abstraction defined by the standard are:

- Algorithmic level
- Programmer's view
- Programmer's view + timing
- Cycle accurate level
- Register transfer level

In addition to exploring questions that relate to the hardware architecture of the device, explorations can be made into the device firmware and application software spaces as well. Architectural simulations can run fast enough to give software engineers the first taste of the device that they will be programming for. This can allow software engineering activities to begin much earlier in the design cycle. Traditionally, these activities begin fairly late. By starting earlier in the design cycle, software architects can identify key issues in the hardware that make software construction difficult and request that those issues be resolved while it is still relatively easy.

Architectural modeling provides a fast simulation platform for modeling architectural and software experiments by abstracting out all elements of the design that are unnecessary for these studies. Given the key parameters that define a bus's bandwidth, the architectural simulation can model traffic over the bus without taking into account how the bus will be implemented in actual hardware, or the low-level details of the drivers that will feed the bus. The simulator no longer has to spend processor time modeling unnecessary information and can run faster.

There are a number of programming languages that can be used for architectural modeling. Perhaps the most popular of these is SystemC. SystemC is built on top of C++ and encapsulates certain design concepts such as time and signals. By using SystemC, the architect does not need to concern themselves with modeling these low-level concepts and can simply make use of the constructs that are already defined by the language.

Other languages have been used for architectural modeling over the years including Verilog, VHDL, plain C/C++, and e. In addition to these languages, some architectural studies are carried out using readily available tools such as spreadsheets.

What Metrics to Track

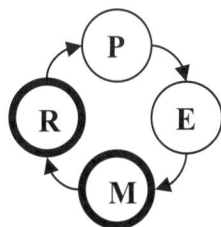

Because it is a form of verification, the goals of architectural modeling activities should be tracked in a verification plan. The main goal of architectural modeling is to determine if the selected architecture for the device under design will be able to provide enough functionality for the desired feature set.

The verification plan essentially lays out a list of metrics that must be measured to satisfy each architectural study. By tracking these metrics to completion, we are ensured that the appropriate metrics have been considered and that the proper architectural trade-offs have been made.

Architectural modeling metrics will consist mostly of functional coverage metrics. Functional coverage provides information about what scenarios have been exercised in the modeled device.

If there are multiple architectures that are being evaluated, then the metrics for each architecture should be tracked independently. Using this information, architectural decisions can be re-examined later in the project, or as the project is modified for further revisions.

Architectural models often contain complete or almost complete descriptions of the final device's behavior and correct operation. As such, they can be used as reference models for the later simulation and emulation stages of verification. One key historical metric that should be tracked is the number of modeled behaviors. The methodology for using each modeled behavior should be documented.

The functional coverage of architectural modeling should track various architectural concerns including, but not limited to:

- Performance
- Bus traffic bandwidth
- Algorithm studies and selections
- Memory size selections

Any concern that is to be resolved by an architectural study should be tracked by functional coverage to ensure that the study was in fact completed. Tracking the coverage is not enough however. In addition to ensuring that the study was completed, we need to track that the necessary architectural decisions were made using the resulting data and how those decisions were made. Consequently, another artifact or metric that should be tracked is documentation of decision evaluations with specific links back to the recorded functional coverage metrics and results.

Functional coverage metrics should trend toward 100%. Architectural studies should be re-evaluated after every change of the device's architecture. For example, when a device peripheral is added or removed, all architectural study simulations related to the bus that the peripheral uses should be rerun.

How Architectural Modeling can be Leveraged
The stimulus, coverage and correctness models developed for architectural modeling can be reused throughout the project. As mentioned in Chapter 4, the correctness models used for architectural modeling can be reused for checking the correctness of hardware models as well. Transaction-based stimulus can be reused in a properly architected verification environment as well. Transaction-level modeling also allows architectural coverage definitions to be reused.

By reusing checks and coverage from the architectural stage, the verification team can perform a sanity check of the architectural assumptions as they are gradually translated into actual hardware. By performing these sanity checks after each block is completed and after each new level of integration, catastrophes caused by incorrect architectural assumptions can be avoided through early detection.

The architectural model is often used by software developers to develop the first cut of the system's firmware. Because of their high level of abstraction, these models can run fast enough to provide useful results for software developers.

Assertion-Based Verification

Why Use Assertions?

The sooner bugs are spotted in the design, the cheaper they are to fix. It has been shown that the cost of fixing a bug in a design increases exponentially with time.

For example, look at the design shown in Figure 5.6. It can be seen that the DMA block moves data for the DSP, LCD driver, and the PWM DAC. The DSP can control the operation of the DMA block. If bugs are found in the LCD driver at the chip integration level, we have to investigate not only the LCD driver, but also the DMA block, and the DSP.

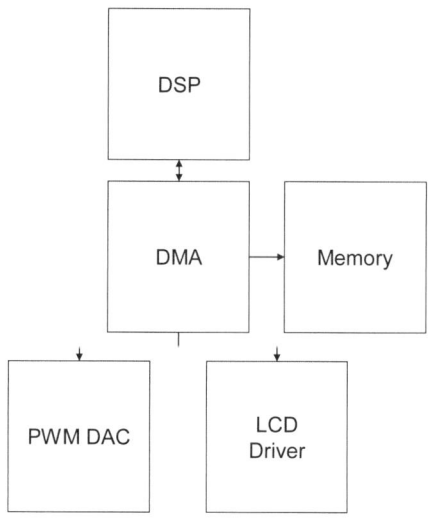

Figure 5.6 Assertion Example

First we would verify that the LCD driver was doing the correct operations based on the control data it received. We would then check the DSP to make sure that it was correctly driving the DMA engine. Once again, we find that there are no problems. Finally, by the process of elimination, we check the DMA engine and find the bug. So, after checking three hardware blocks, and one piece of DSP software, we finally arrive at the root cause of the bug.

In contrast, if we had caught the bug at the block level, we would have investigated only one hardware block. With assertion-based verification that's exactly what we'll do.

How Assertions Work

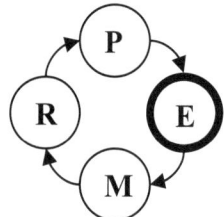

Assertions can be easily used to test block-level functionality at the block level where it is still easy to fix. Assertions are simple Boolean or temporal checks that are easy for both design and verification engineers to write. An example of something a Boolean assertion would check is:

```
The FIFO write signal can never be
asserted when the FIFO full signal
is asserted.
```

Temporal assertions are slightly more complex because they can utilize timing structures. An example of something a temporal assertion might be used to check is:

```
When the request signal is asserted,
the grant signal must be asserted
between 10 and 15 clock cycles
later.
```

Assertions can be written in design languages such as Verilog, VHDL, and SystemVerilog. That makes them easy for designers to write. They can also be written in more verification centric languages such as *e*, property specification language (PSL), and

SystemVerilog Assertions (SVA) making them easily accessible to verification engineers as well.

There are two aspects to any verification issue. They are controllability and observability. As was shown above, assertions and block-level verification certainly improve design visibility and subsequently reduce debug time. However, the design must still be exercised via testbenches and testcases to actually verify the functionality that the assertions were designed to check. That means that someone still has to write some kind of a testbench and testcase to fully verify the device even at the block level.

However, there is a class of tool that can eliminate even testbench creation and testcase writing. It makes block-level assertion-based verification even more effective. These tools are typically called formal verification tools. They use formal proof engines to prove that a given block of hardware can never violate an assertion. These formal tools work best on small blocks of hardware because the complexity of the resulting proofs that must be solved increases exponentially with design size. They also work best on blocks that contain mostly control logic such as state machines as opposed to blocks that are used to transform data such as multipliers.

These formal assertion checkers can be used even before there is a testbench available for the device. As soon as there is a design module that contains viable design code, the designer can start to prove their assertions. Think of the value here. The designer is writing checks for their own blocks and receiving instantaneous feedback about the viability of their design choices. They're catching bugs in the design almost the instant that they are created. Actually, if they use the planning techniques outlined in Part III, designers can catch many bugs *the instant* they are created!

In order for formal assertion engines to solve their proofs, designers need to tell the engines what behavior is allowed at the inputs of their device and what behavior will never take place. The formal engines will then use this information to simplify the mathematical problem that they have to solve.

For example, in the request grant scenario described above, the block may have been designed with the assumption that no two requesters would be allowed to request the block at the same time. The designer would write an assertion called a property or assumption at this level to convey this information to the formal engine. It would read something like:

```
Only one request signal may be
asserted at a time.
```

As we begin to integrate blocks of the design, our formal engines will quickly be outstripped by the designs complexity. So, is that all we get out of our formal processes?

Not quite. The properties that bounded the input space in formal analysis are in fact rules that the integrated blocks must follow at the unit, chip and system integration levels of verification. By telling the simulation tool to treat the formal properties as dynamic assertions, we have automatically added a number of useful checks to our verification environments without our verification engineers having to write one line of additional code.

The picture gets a little brighter still. Remember that complex debug cycle to determine that the DMA block was actually at fault? Well, we're bound to still get a few of those. As we move up in integration levels, we typically turn off assertions to enhance simulation and emulation performance. Now, when we find the same bug our first step will be to turn on all our block-level assertions. Sure enough, we find when we resimulate that the DSP software utilized the DMA in a manner that wasn't expected and forced a piece of the hardware to violate one of the initial design assumptions. Rather than tracking through three hardware blocks and a software block, our designer merely turned his assertions back on and was led straight to the problem.

To summarize, we have our design engineers checking device behavior very early in the design cycle when it's very cheap to fix bugs. They don't have to wait for verification engineers to create testbenches or testcases for them. And, in some cases, not only can the design engineer check functionality, they can prove that the functionality will always work based on their assumptions. Those same assumptions can serve as additional checks that the various blocks of the design play well together when the design is promoted to higher levels of integration. Finally, even at the chip and system level where we hope all is well in our individual blocks, if it's not we can re-enable our assertions and use them to provide precious debug information.

What Metrics to Track

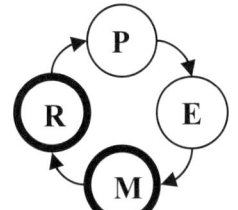

Now we know what value assertions can provide, but how do we track that they are being implemented efficiently during the project? There are a few objective and heuristic metrics that can help us out.

During verification planning, the designers identify pieces of functionality that can be verified using formal analysis. We can automatically track the status of those defined assertions. This information can be used in conjunction with revision control information in two ways.

First, we should see the assertions move to a proven state at roughly the same rate that we see the number of lines of code for the given design module increase. If there are spikes in the size of the design module vs. the number of proven assertions, that doesn't necessarily mean there is a problem. Development style varies from engineer to engineer, and some engineers like to code larger portions of their design before starting to test. However, if we see the design block declared complete and there are still no proven assertions as defined in the verification plan, then there is a red flag that bears checking.

Second, every assertion of a given block should be reproven every time the block is modified. If the assertions aren't reproven, bugs caused by the code modification can slip through.

As we begin to move into simulation where dynamic assertions will be used, there is another metric to check. Dynamic assertions are only valuable if the functionality of the device that the assertion checks is actually exercised. All assertions can be divided into a qualifying portion of the assertion, and the actual check itself. Let's look at our temporal assertion example again.

```
When the request signal is asserted,
the grant signal must be asserted
between 10 and 15 clock cycles later.
```

The qualifying portion of the assertion is that the request signal was asserted. The check is that given the qualifying condition, the grant signal must be asserted properly.

We need to actually track that the assertion's qualifier was exercised. Assertion coverage tools can be used to automatically track this information. If the qualifier is never exercised by the regression suites, then they need to be enhanced to create the desired stimulus. There is one caveat here. It might be determined that the assertion was never checked because the device in fact does not operate in a manner that ever stimulates such behavior.

Finally, there is a heuristic metric that is of most use in projects where verification planning is not used. That metric is simply the proportion of lines of assertion code in each design module vs. the number of lines of design code. It is similar in nature to the old software heuristic of requiring that twenty percent of the lines in a given software module should be comments documenting the code. This metric provides a rough feel for whether or not engineers are properly utilizing the available assertion-based verification tools, although it is by no means as objective as the other metrics discussed here.

Who Utilizes Assertions

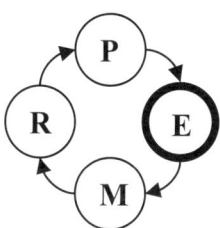

From a tactical standpoint, designers make the most effective use of assertions. They can use their intimate knowledge of the underlying design to quickly code these small checks. The rewards, that some portions of the design are "proven" correct and need not be further verified and that bugs are found while they are still easy to fix, provide suitable motivation to utilize assertions.

Verification engineers utilize assertions at the boundaries of design blocks to reap tactical rewards on the current project and to garner large strategic rewards in future projects. These verification-based assertions check the rules for communication between design blocks. When verification engineers write these assertions, they are creating an independent contract that the disparate authors of adjacent, communicating design blocks must adhere to. Tactically speaking, integration of design blocks becomes much easier as communication issues are caught in the initial stages of integration. Because these verification assertions are independent of the design implementation, they can be reused from project to project, either traveling with a given design block, or being used as an implementation contract when new design blocks with the same communication protocols are created.

How Assertions are Leveraged

Design-based assertions can be tactically leveraged beyond block-based verification. At the integration, chip and system levels, these assertions can be reactivated to provide valuable information that can greatly enhance debug speed.

Verification or protocol assertions can be leveraged strategically across projects. These assertions can be reused either with existing design IP as it moves from project to project, or on new design IP that must conform to the given protocol.

Formal-based verification requires the definition of assertions that define boundary assumptions. These assertions define the legal input behavior for a block. They can be reused as dynamic assertions when the block that was formally verified is integrated with the adjacent block that drives its inputs. In this manner, it is easy to find outputs from the adjacent block that incorrectly drive the originally verified block.

Simulation-Based Verification

Why to Use Dynamic Simulation

Formal assertion-based verification is great for proving that block-level control logic works. Assertions also allow us to define complex Boolean and temporal checks easily. However, once these checks have been defined, their qualifying conditions must be created. If these assertions are not formally proven we must use dynamic simulation or emulation techniques.

Simulation is a software technology that allows engineers to fully model the behavior of a semiconductor device. The device can be modeled at a fraction of its actual speed and the input and output waveforms are analyzed for proper behavior.

By using simulation-based testbenches, verification engineers can create complex scenarios to test the functionality of various features of the device. Using high-level behavioral programming languages, verification engineers can express these complex scenarios in a concise manner.

Several software debug and analysis tools have been developed for the simulation arena. These allow engineers to view signals within the simulated device as if they had the actual device and were testing it on a workbench.

How Simulation Works

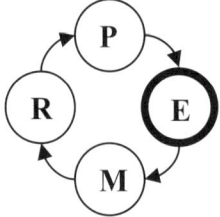

Simulation-based verification consists of three steps. The designer develops a model of the device in an HDL language such as Verilog, VHDL, or SystemVerilog. In parallel, the verification engineer develops a testbench for the device. The testbench is used to instantiate the device (create a model of the device in the computer's memory), and then drive the device's inputs and read the device's outputs.

There are two broad classes of simulation testcases. These are usually denoted as directed and random testing. Directed testing is carried out in much the same fashion that software is written. The engineer sits down and writes a procedural description of a test scenario that should be run on the DUT. With a directed testcase, the engineer can create exactly the test scenario that is desired. For simple testcases, directed testing is perfect. The engineer can easily create a scenario that simulates the device and checks for proper behavior. However, when complex timing relationships are involved, writing directed testcases can become quite difficult. Once a testcase is created, a subsequent change in the design timing can render the testcase useless. In short, an engineer can create exactly the scenario they want with a directed testcase, but possibly not for very long, and they *only* get the scenario they want.

Random testing, or as it is more properly called, constrained random testing makes use of randomness to automatically create desired scenarios and to create scenarios that are legal, but that were not specified in the verification plan. There are two primary values to this type of testing.

First, the complex testcases mentioned above can actually be generated automatically. The engineer creates a test environment that randomly drives stimulus to the DUT. They also create mechanisms that detect the desired scenarios. These monitors are called functional coverage groups. The engineer then runs multiple simulations using the randomly created stimuli and checks to see that all the desired scenarios were created. If there are scenarios that are not created, the engineer constrains the stimulus to create more favorable stimulus.

The second reason for using constrained random techniques has to do with the rapidly increasing complexity of today's designs. As the number of features of a device increases, the number of operational combinations of the device increases exponentially. It has become impossible to exhaustively test all the states of the device. Verification efforts are usually constrained (no pun intended) to the most important configurations of the device. Random testing will test the device in manners that are legal (via constraints), but that were not originally specified in the verification plan. This can lead to the detection of bugs in the DUT where no one ever thought to look for them.

What Metrics to Track

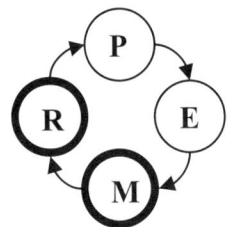

There are two key metrics used to track the progress of dynamic simulation activities. These are code coverage and functional coverage. Code coverage measures how many times each line of a design description has been executed. Functional coverage is defined by the verification engineer and measures how many times a specified scenario has been executed.

By itself, code coverage can tell the design engineer if all their code has been exercised. This is certainly a necessary condition for good verification, but by no means a sufficient one. The fact that every line of code has been executed does not imply that all the functionality of the device has been exercised.

But, consider code coverage data in conjunction with functional coverage data. This provides a whole new level of confidence. There are four possible scenarios that are illustrated below.

If we have high-functional coverage and high-code coverage, then *probably* life is pretty good. There could be one problem here, but it is unlikely. The design engineer could have *not* implemented exactly the same portions of design code that the verification designer implemented no functional coverage for. Another way to say it might be that the design and verification team were not aware that new features had been added to the device. This unlikely (?) event can be prevented by tracking both these metrics to the verification

5 Verification Technologies 73

plan and making sure that the team in charge of feature definition has visibility into that document.

What about the other cases? What if code coverage is high, but functional coverage is low. This could indicate that stimulus for all the functionality of the device had not been created. It could also indicate that portions of the design corresponding to the missing functional coverage have not been implemented yet.

What if the code coverage is low, but the functional coverage is high? It could be that portions of the functional coverage are not yet implemented. It could also indicate that there are design structures that offer no actual functionality. Perhaps a feature was cut that the verification team was aware of, but the feature has not yet been removed from the device.

By taking two metrics that were readily available, we went from a simple good bad analysis (code coverage alone) to an analysis that begins to shed light on what is actually going on.

Mixed-Signal Verification

Why to Use Mixed-Signal Verification
It's an analog world. The real-world (ignoring quantum mechanics and string theory) produces a rich continuum of values. The digital world operates on ones and zeroes. In order for interesting digital applications to work on real-world data, the two must meet. That's where mixed signal simulation comes into play.

Planning for Mixed-Signal Verification

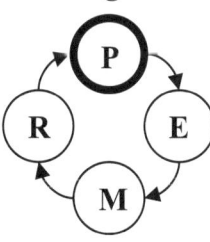

From the digital point of view, planning for mixed-signal verification is exactly the same. The analog portion of the problem adds a few new twists however. Now in addition to considering what a device feature does, the engineering team must consider the environmental conditions the device itself lives in. Planning must take into account the analog behavior of the device at different operating temperatures. In addition, the plan should also account for the different component description "decks" that describe the different corners of the silicon manufacturing process.

Digital blocks are frequently not only the recipients of data from analog blocks, but also control how the analog block functions. For an example, see Figure 5.7. Here an analog preamp feeds an analog to digital converter that supplies input data for the rest of the digital domain.

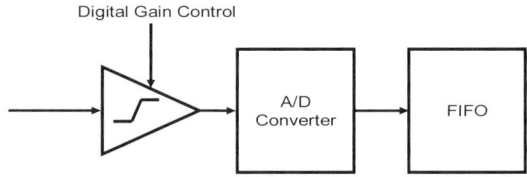

Figure 5.7 Digital Control of the Analog Domain

Note that the gain of the preamp is determined by a control signal supplied by the digital domain. When planning verification it is important to check all combinations of the digital and analog domains that affect each other. For example, the environment shown above should verify the different operating temperatures, in combination with the various gains that can be provided to the preamp block.

How Mixed-Signal Verification Works

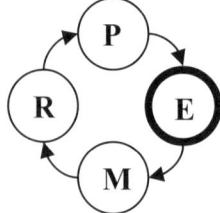

Analog simulation can be performed in much the same manner as digital simulation. Engineers construct a description of the circuit based on the actual components within the circuit instead of the synthesizable behavioral constructs used for digital logic design. These

5 Verification Technologies

circuit models are then instantiated into an analog simulator where the actual operation of the circuit can be simulated in a variety of operating conditions with a variety of input stimulus.

Depending on the level of abstraction, analog simulation is much slower than logic simulation. Historically this has resulted in fewer testcases being run and little coverification between the analog and digital domains.

However, with increasing design complexity both in the analog and digital domain, it has become essential to move to a metric-driven coverification approach.

For more information on how analog/digital coverification is implemented, see the corresponding chapter in Part III.

What Metrics to Track

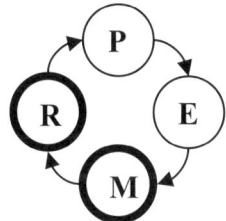

Trends in mixed-signal verification are tracked with the same techniques used for tracking architectural or simulation status. One key difference is making sure to keep analog metrics such as manufacturing deck and operating temperature in mind.

Acceleration/Emulation-Based Verification

Why to Use the Technology
As more of the various parts of a design are integrated, simulations run more slowly. There are speed and memory space limitations that are imposed on simulators by large designs. Ultimately, simulator performance can slow to such a level as to be useless. For example, when simulating the complete chip for an MP3 decoder, it can take up to three days just to decode a single frame of an MP3. One MP3 frame is not long enough to be heard by the human ear.

Planning for Acceleration and Emulation

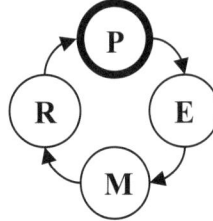

Acceleration and emulation are typically used for system-level verification. With that in mind, the planning process is tilted toward the chip-level integration engineers, the firmware engineers and the application engineers. From a chip-level integration engineer's point of view it will be important to make sure that the chip as a whole comes to life. The chip should behave correctly coming out of reset and the various blocks should be able to communicate. There is less emphasis on the correct behavior of each feature of each block of the device because it is assumed that this has been verified in an earlier stage of verification using dynamic simulation and/or assertions.

The firmware engineer will need to check the code that will execute on the device and provide basic functionality for the higher level application code. Firmware engineers sit squarely between the hardware design and software application worlds. Planning from a firmware point of view will focus on such issues as whether each firmware module has been executed (software code coverage), and whether certain hardware corner cases were encountered during the execution of the firmware (did each type of allowable interrupt occur during peripheral initialization?).

Application engineers are interested in verifying that the device works with real-world application code in the target system that customers will use. These engineers will be interested in exercising the device using in-circuit-emulation (ICE) to attach it to real-world target systems. For applications, verification planning will be focused on tracking that each application was run in each legal configuration of the device with a variety of real-world system topographies.

How the Technology Works

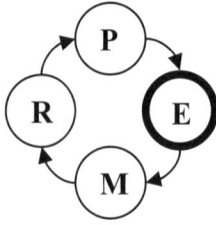

When simulation performance is no longer acceptable, many organizations turn to emulators and accelerators. Emulators and accelerators are used to test a device in much the same way that simulators are. Engineers download the same device definition that was used for the simulator

5 Verification Technologies

onto the emulator or accelerator. While simulators are software applications, emulators and accelerators are specialized hardware solutions. This specialization offers great performance gains. Typical emulators and simulators run anywhere from ten to one thousand times faster than a simulator.

Emulators and accelerators are similar in their operation and performance benefits. There is one key difference between the two however. Emulators are best suited for ICE. Using ICE, the emulator can actually plug into the target system that the real device will be used with. This allows engineers to verify the emulated device as part of the real-world system that the production device will eventually live in. Accelerators are typically somewhat easier to use and are best suited to using the same testbench that was used in simulation with no connections outside the accelerator hardware. These are only the sweet spots. Of course, emulators can be used in a targetless environment and accelerators can be used for ICE.

This higher speed does come with a few limitations. Emulators and accelerators are much more expensive than software simulators. They typically require more initial setup effort than software simulators. It is also typically harder to extract debug information from these tools.

Who Utilizes the Technology?

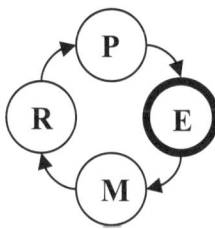

The performance provided by these solutions makes them ideally suited to system-level verification. They are used to verify entire chips both with and without their target firmware and application software. Emulators and accelerators often offer the first chance to verify the hardware/software system as a whole. Because these systems offer real-world execution speeds and debug access, they are often used by software teams after being setup by the design and verification teams.

What Metrics to Track

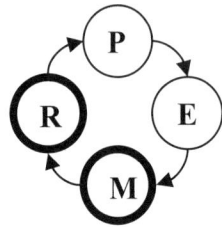

As with the other verification technologies, coverage is one of the key metrics to track when using emulation and acceleration. The coverage metrics used change a bit on these platforms however. Code coverage is no longer used. Functional coverage is used to track system-level concerns when using emulators and accelerators. Concerns such as which software methods have been called are tracked. While signal-level functional coverage can still be used, there are performance vs. applicability trade-offs to be considered.

How the Technology can be Leveraged

Testbenches used for simulators can be reused in accelerators and emulators if properly designed. Emulators and accelerators work only with synthesized designs. Care must be taken to implement the testbench in a synthesizable manner if it is intended to be reused at this level. Assertions can also be reused in emulators and accelerators to provide valuable debug information in the event of a detected failure.

Summary

In this chapter we have looked at several of the most popular verification engines and given a brief overview of how each of them fits into an MPA framework.

Part II
Managing the Verification Process

Preface to Part II

In Part II we describe in more detail how to apply a metric-driven methodology to our every day processes. We move down a level of abstraction form Part I and discuss how to actually implement metric-driven processes. We'll discuss implementing the "container processes." Container processes are the processes that apply regardless of what underlying verification methodologies we are using; processes such as regression management and revision control. Working within the plan, execute, measure, and respond framework described throughout Part I, we'll first describe in detail how to plan verification projects.

Next, we'll look at a layered methodology for capturing metrics independent of the source of the metrics in Chapter 7. Presenting a methodology to capture metrics breaks away from the plan–execute–measure flow, but it makes sense in the context. Even as we start our execution engines, we need to capture metrics, not only metrics returned from our executions engines, but also metrics about how and when they are used. We actually need to have our metric capture apparatus in place, functional, and *visible* before the execution phase of the project begins.

In Chapters 8–10 we'll discuss the "container" processes of regression management, revision control and debug.

Chapter 6
Verification Planning

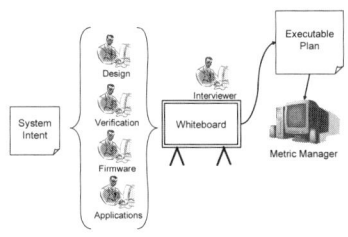

Introduction

Two little boys sat on the front porch of their house one morning. One looked at the other and said:

"We need to learn how to swear today."

"OK," said the second little boy, "I'll say damn and you say hell."

After awhile, their mother came out and said, "What would you boys like for breakfast?"

"Well damn! I think I'll have some cereal!" said the first little boy.

His mother grabbed him by the back of the neck and dragged him into the house. For several minutes after that, all the other little boy could hear was screaming.

Soon the mother returned looked at the second little boy and said, "What do you want for breakfast?"

The little boy replied, "I don't know, but I sure as hell don't want any cereal!"

Getting chip verification right is all about understanding each stakeholder's translation of product intent correctly. It's all about understanding what was originally intended for the chip and about matching everyone's understanding. That's why the verification planning methodology described here is collaborative. One of the foundations of this methodology is to gather EVERYONE for the planning process. That means that the design engineers, the verification engineers, firmware engineering, apps, and management all need to spend some time in the same room together.

Verification planning is the lynch-pin of everything else in the verification process. This is what will enable us to catch bugs earlier. This is how we will set our objective goals so we know when we are finished. This will be the basis of our automated status reporting that will be visible to all stakeholders in the project. This is *not* where we will spend a lot of time! We need to do just enough planning to make sure that the project is constructive and then get started on the real work, the *project*!

We're starting with metric-based planning for a few reasons. First it will enable us to make the most effective use of automation tools moving forward. By planning and implementing our coverage first, we immediately know what is being accomplished by each of our activities, methodologies and tools. Without the coverage in place, we can turn these tools on, but we have no way of knowing what they have done.

Second, metrics makes the project more interesting, safer, and more secure. Ever watched a ball game and not cared about the score? Metrics give us visibility into what is being accomplished on the project at each given instant. It answers questions like:

- How far along is the verification of the serial block?
- Has anyone used the IIS block in stereo mode?
- Have we tried back to back cycles on each of the AHB busses?

The project is more interesting to all the stakeholders involved because they can *see* what is happening in their area of concern at any given time. We hook the interest of everyone involved in the project and keep it. There's something to watch all the time. Stakeholders don't wander off, and work on other tasks for days or weeks at a time and come back only to be amazed that the wrong things or nothing has happened in their area of concern.

The project becomes a safer place. By properly designing and parsing our coverage into chunks, we completely avoid that status meeting phenomenon that we're all so familiar with. "Well, it's coming along and I'll have it done soon" repeated week after week with no real evidence of what is coming along or how. We have a measurement every day of what has been accomplished. The project is safer for leaders because they're not at the mercy of ambiguous status. It's safer for the team members, because if there is an issue that is inhibiting progress, it becomes obvious earlier, and the entire team can work as a *team* to resolve the issue and move on.

Finally, the project consistently maintains a higher quality. Our planning and the visibility it provides will remove opportunities for undetected failures in the device or the execution of verification. The objective measurements we're making project-wide visible will focus the proper stakeholders on the proper aspects of the project so there are no surprise requests for verification in the ninth hour. With these concerns removed, the verification and design teams can concentrate on the job at hand.

Chapter Overview

In this chapter we outline the process of verification planning. There are two purposes for verification planning. The first is to ensure that everyone involved in the design project (design, verification, firmware, etc.) have the same interpretation of the product intent.

The second purpose is to capture in a single document, the concerns of each stakeholder with respect to design verification. First, we'll capture the aspects of each system feature that concerns the stakeholder. Then we'll define how to automatically measure that their concern has been answered in a satisfactory manner. For example, concerns in Table 6.1 might arise during a verification planning session for a DMA block.

Table 6.1 Concerns and Measurements

Stakeholder	Concern	How Measured?
Application engineer	The LCD controller should receive 60% of all available bandwidth when multiple requestors contend for the DMA controller.	Use a functional cover group to show that multiple requestors contended for the DMA controller. Use a checker trigger to show that the bandwidth requirement was checked.
Design engineer	The input FIFO of the DMA controller generates an interrupt on a write request when it is full.	Use an assertion trigger to show that a write request arrived when the input FIFO was full.
Design engineer	The DMA controller code should be sufficiently exercised.	Measure 100% code coverage for the DMA controller module.
Verification engineer	The DMA controller should move blocks of memory between 1 and 1024 bytes between different memory addresses.	Use functional coverage to show that every transfer size has been exercised within a suitable range of input and output addresses. Use a check trigger to show that each transfer was checked.
Firmware engineer	The DMA block should copy the interrupt handling code from the embedded ROM to the instruction cache.	Use functional coverage to show that transfers of the appropriate size from the ROM to the instruction cache have been executed.

6 Verification Planning

Using these captured concerns and measurements, we will create an executable verification plan. The purpose of this plan will be to automatically track each measurement (or metric) during the life of the project. The results of each measurement will be automatically annotated into the plan so that stakeholders can continuously track the status of their concerns as the project progresses.

To accomplish the first goal of verification planning (product intent interpretation convergence), we'll use a collaborative brainstorming process. All the stakeholders will meet to discuss their verification concerns and a facilitator will capture these concerns in a document that will become the executable verification plan discussed above. Figure 6.1 graphically outlines the verification planning process.

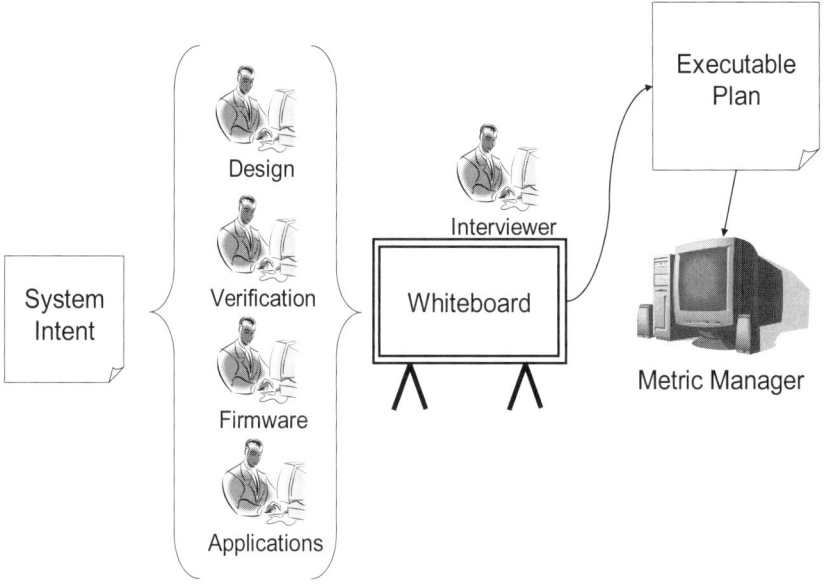

Figure 6.1 The Verification Planning Process

This chapter will describe how to perform a successful verification planning session. It will outline how to best organize and execute the collaborative brainstorming/interview sessions required. We will discuss the process from several points of view. First, we'll look at the generic interview process. Next, we'll look at how each stakeholder can be interviewed. We'll focus on pertinent questions to ask each type of stakeholder, how each type of stakeholder can add value to the process, and how each stakeholder's concerns might be converted into automated measurements as mentioned above. Finally, we'll take one last look at the process from the perspective of each design integration level. We'll explore how the process should be performed for different phases of the project such as block-level design, unit integration and chip-level integration.

This chapter will contain the following sections:

- Overview
- The planning brainstorming/interview process
- Identifying system features
- Identifying system feature attributes
- Planning with respect to verification
- Planning with respect to design
- Planning with respect to software
- Planning with respect to architecture
- Planning for block-level design
- Planning for unit-level integration
- Planning for chip-level integration
- Planning for system-level integration
- Decorating the plan: goals, weights, and milestones

Verification Planning

Verification planning will be done in several different stages and in several different perspectives over the course of the project. It is also a very iterative "living" process, so we won't worry about getting it perfect the first time. The stages of verification planning are:

Coverage Planning: "What to Verify and How Do We Know It Happened?"

Checking Planning: "How Do We Know It Worked?"

6 Verification Planning

Stimulus Planning: "How Do We Make It Happen?"

The coverage planning stage is performed first because the other two stages can and should be derived from the coverage planning.

Determining what to verify is the most important stage and the only stage that we'll discuss here. During the verification planning sessions described here, the participants should be particularly careful not to get wrapped up in how to check the correctness of a feature. The verification technology that will be used may be defined, but no effort should be put into determining exactly how to do the checking or provide the stimulus. We'll do that later. During verification planning we want the focus to be purely on what should be verified.

The Planning/Brainstorming Process

The verification planning session follows the familiar brainstorming format. As with all brainstorming, there are a few ground rules. There is one moderator and one note taker. It is the job of the moderator to keep the session on track according to the agreed upon rules. The note taker is to collect all output from the meeting so that when participants wrap-up the session, the first draft of the verification plan exists and can be distributed.

The purpose of the brainstorming session is to capture all the features of the given device and also to capture each stakeholder's concerns about that feature. The rules are:

- The majority of brainstorming should be done on a whiteboard and the information should be derived from the participants' experience. The specification is only to be used to clarify issues where there is no agreement.
- The participants are to discuss what the feature *does*, what to *measure* about the feature, and how to detect *when* the feature has been exercised. They are not to discuss how to check the feature. Discussions of how to do checking can quickly become too involved and sidetrack the discussion. Remember, all we want to do here is capture all the features and each stakeholder's concerns as quickly as possible. We'll architect the verification environment later.

All the stakeholders should be present for at least a portion if not all of the session. Participants should be reminded that the ROI on their time in the planning session will be huge! Among other great benefits, this is their get out of jail free card for the endless series of status meetings that plague the last stages of most design projects. The planning session should consist of participants from:

- Verification
- Design
- Architecture
- Firmware
- System Applications
- Product Definition

One participant that is familiar with the verification planning process should be designated as the moderator. It will be the moderator's responsibility to maintain the flow of the session. Another participant or a rotating queue of participants should be appointed as the session's stenographers. In reality, this person does not need to have a command of the design. An administrative assistant could be utilized here. The key consideration is that this person needs to be able to capture all concerns in our planning template without interrupting the flow of the session. The stenographer will be constantly redirected within the document and should keep in mind that they are for the most part capturing information, and, as with all brainstorming, the participants should not be inhibited from remapping the document.

In addition to the participants, a small modicum of equipment is required as well. A whiteboard is essential. Obviously, the stenographer will need a laptop or other device to record the document on. In addition, the team may find it useful to have a digital camera to capture whiteboard shots with. Access to the specification of the device is helpful. However, the specification document is by no means meant to be used as a key input to the process. This is a communications and brainstorming process, not a document review! Finally, in order to facilitate recording milestones, access to the current version of the project schedule is helpful.

6 Verification Planning

The Planning Session

The output of a planning session will be a feature-oriented verification plan. To this end, it is the job of the moderator to lead the group to discuss each feature of the given device.

A feature is something that the device under verification does. A feature is a verb or activity. Examples of features are:

- A device *provides* data to a read request.
- A device *consumes* data from a write request.
- A device *routes* a transaction from one port to another.
- A device *decodes* an MP3 stream and *provides* audio output

The brainstorming process begins with an interview of the designer responsible for the given design unit of the system that is to be verified. The designer first draws a block diagram of the design under verification (DUV) on the whiteboard being careful to include all input and output ports and any internal architectural features of the device (such as FIFOs, state machines etc.) that they feel are important.

The moderator will begin by asking the designer questions about the features of the design. The discussion usually begins with the various interfaces to the design. The moderator should ask questions regarding how the device is accessed, and how the device outputs data. Questions during this stage might be:

- "What protocol is used to configure the device?"
- "Is the entire protocol used, or are there transactions types that the device legally ignores such as burst transactions?"

The moderator is trying to create conversations about what communication protocols are used by the device and how they are utilized. The end goal is to determine what interface functionality needs to be exercised and checked.

Once the basic interface features are established, the moderator will focus on corner cases. Questions such as the following will arise:

- "What is the correct behavior for the device if the serial input port is driven before the device is configured?"
- "What happens to an output transaction if a reset is asserted before the transaction is complete?"

As the discussion moves from feature to feature, the moderator encourages each participant to comment on what significance that feature holds for them. For interface features, the designers of adjacent blocks might have unique concerns about what protocol transactions are to be implemented. Meanwhile, the system architect may be interested in the required bandwidth for the device. It is this discussion of concerns that provides convergence of the translation of product intent of the various stakeholders.

Once the block interface features have been defined, the discussion proceeds into the other features implemented by the design. During this portion of the interview, the conversation focuses on what the design *does*. The moderator first asks the designer to describe a given feature. The moderator then opens up the floor for discussion of the feature. The moderator should ask questions that inspire each participant to contribute their own understanding of the feature, how they intend to use that feature, and what the most important attributes of the feature are to them.

The important attributes that are identified for each feature lead us directly to the metrics that we will need to capture from our verification engines. When a facilitator begins to capture feature attributes, there are three essential questions that must be asked:

- "*What* is important to measure about this attribute?"
- "How do we know *when* this feature has been utilized so that we can measure the attribute?"
- "*How* do we measure this attribute"

With these three basic questions, the moderator now has the information necessary to create the metric definitions that will be used to observe the progress of the verification of this feature. The interview is a simple process of asking each stakeholder for each feature, "what, when, and how."

Let's take a look at a sample interview:

Facilitator: What's another feature of the device's input interface?

Designer: It can be operated in burst mode.

Facilitator: What's important to measure about the burst mode?

Designer: We need to make sure that all three burst sizes have been tested. They are 2, 4, and 8 word bursts.

Facilitator: When can we tell that a burst is complete and how do we measure the burst size?

Designer: When the burst_n signal goes high, a burst transaction has completed and the size can be read from the bsize bus.

Facilitator: How can we best measure and record this attribute?

Verification engineer: I can capture that information in a functional cover group.

As the questions were being asked and answered, the session's note taker would have captured the information in a document that looks like Figure 6.2.

Device Input Interface
Burst Transactions

The device can accept burst transactions in sizes of 2, 4, and 8 words. When the burst_n signal transitions to high, a burst transaction has been completed. By sampling the bsize bus, the burst size can be determined.

cover: /sys/interfaces/dma_in/burst_trans

Figure 6.2 Verification Plan Excerpt

The "cover" directive will tell the MPA tool which cover group in the verification environment provides the coverage metric that will be used to measure progress on the verification of this feature.

That's it! It's that simple! Now, let's look at the planning interview in a little more detail with respect to each of the stakeholders. The following sections will point out the value that each stakeholder can contribute to the planning process. They will also point out key considerations to keep in mind when interviewing these stakeholders.

Planning with Verification Engineers
The verification engineer will tend to look at the device based on its black-box functionality. They will be very interested in how the device can be configured and what operations (features) the device will exercise in each configuration. The verification engineer will tend to be very thorough in pointing out all the features of a device. This thoroughness, combined with input from the applications and firmware engineers, can be used to define a set of features that must be verified for the device to be declared usable.

Verification engineers should be very adept at attribute specification. There is a tendency among verification engineers to immediately bore down to the implementation of checkers for the various features. This tendency must be avoided. The checker design will come soon enough. For the planning session remember that the main goal is to capture the metrics that will be used to measure the progress of the project.

Planning with Respect to Design
Design engineers are very concerned with how their block, subsystem, or chip implements various features of the design. Design engineers are the best source for information regarding what white-box coverage and checks should be included in the verification environment. Many of the concerns raised by design engineers can be addressed by assertions that they themselves can place in the design. This is one place where the line between coverage and checking emphasized in this process begins to blur. Using the example from the overview, a design engineer may identify a feature:

"When the input FIFO is full an interrupt should be generated."

This feature can be easily checked by a design assertion. The trigger of the design assertion, "when the input FIFO is full," can serve as the coverage of this feature. In simple examples like this it is permissible and even advisable to discuss the feature, its coverage and its check all at once.

Design engineers are very specification driven when it comes to the features of the design, as they should be. It is important to balance the design engineer's understanding of a feature with how the users of the feature, such as firmware and applications engineers, intend to make use of it. It is at this juncture that many bugs can be discovered before the design or verification work ever begins.

While a designer of a given block is often the focal point of a verification planning session, don't forget to include the designers of other blocks within the system as planning participants. Designers of blocks that either provide input to the block under consideration or consume output from that same block are of particular value. It is these designers that can point out key attributes of the features that are important to their blocks. By having all the design engineers concerned involved in a single session, major integration bugs can be avoided before the verification environment even exists.

Planning with Respect to Software

Software engineers, both firmware and applications engineers, have one of the best perspectives on how the device will ultimately be used. They contribute planning content that will need to be translated based on the level of integration. At the chip and system level of verification when the entire device is complete, their input can be taken as is and added to the plan. At the block and unit levels, the software engineer and the moderator will have to work a bit more to translate the software engineer's concerns into measurable metrics.

For example, when discussing a single DMA engine within a design, the software engineer may at first feel that their presence at a block-level planning session is unnecessary. It is the job of the moderator to elicit useful information from the software engineer and keep them involved. Questions such as the following will illuminate key attributes that should be covered by objective metrics as the device is verified:

- "How will your software make use of this device?"
- "Is there a configuration mode that your software will use most often for this device?"
- "What is the most important configuration of this device with respect to initial software testing?"
- "What will be the most important configuration mode used by our customers?"
- "What is the default configuration mode that will be used by your software after a restart?"

As the design integration level moves up, the questions to the software engineers become more direct and less inferred. For example, at the system level of verification, the software engineer may very well be the engineer that is interviewed for the planning session as opposed to the design engineer who was interviewed for the earlier block and unit-level planning sessions.

Careful attention to the differences in descriptions of a feature by the software and verification or design engineers can produce very valuable insights that can significantly reduce the amount of effort required to verify a device.

Verification engineers interpret a specification literally without always having a good perspective about how an end-user will eventually use a feature. This can lead to coverage and checkers that are far more detailed than they need to be. An experienced moderator will encourage the software and verification engineers to reconcile thier understandings of a given feature. For example, consider the partial specification of a debug block feature in Figure 6.3.

6 Verification Planning

Floating Point Profiler
The floating point profiler will provide data to the users that allows them to determine how much of their firmware execution time is being spent executing floating point instructions. When a floating point instruction is decoded, a counter will begin to increment on every clock cycle, and when the results for the instruction are stored, the counter will stop incrementing. The counter will be reset to 0 when the profiling feature is enabled and maintain its last stored count once the feature is disabled. The user can determine the number of cycles counted by reading a register named FPTMCNSMP.

Figure 6.3 Floating Point Profiler Feature Specification

Left to their own devices (without any further information from marketing and product definition), a verification engineer might decide that the checker specification given in Figure 6.4 must be implemented.

Floating Point Profiler Checker
The checker will model the floating point profiler. When a floating point instruction is decoded as it is read in on the instruction data bus, the checker's counter will begin to increment on every clock cycle. This incrementing will be stopped when the results of the floating point instruction are stored. If any reads to the FPTMCNSMP register are detected, the contents of the returned read will be compared to the reference model's counter. Any discrepancy between the two counts will be flagged as an error.

Figure 6.4 Verification Checker Specification

There are several difficulties that are inherent in the above feature specification and checker specification. The feature specification states that the counter will begin to increment after a floating point instruction has been decoded by the DUT's internal decoder. The checker is a black-box checker that is detecting floating point instructions by decoding them from the instruction bus. Depending on the pipeline depth of the device's decoder, several cycles of inaccuracy could be inserted here. There are similar concerns for the pipeline depth of the store unit.

The marketing contributor may reveal during the planning process that the user is only interested in statistical data sets accumulated over thousands of floating point instructions. They may also reveal that accuracy within fifty cycles of how long the operation actually took is sufficient for the customer. Depending on the length of the pipelines within the DUT, the verification engineer's job just became much simpler. Rather than trying to determine how to compensate for various pipeline depths within the device, the engineer builds a much simpler checker that merely counts the number of cycles from the inception of the floating point instruction on the instruction data bus to the completion of the instruction when its results are stored back to memory.

Planning with Respect to Architects
Architectural engineers tend to be more concerned with dataflow through the device and the operation of the device as a system than design engineers. Architectural engineers have a more abstract view of the system and are far more concerned with usability and performance issues. The features they identify will have more to do with how the *system* operates than with how a given block operates. As with software engineers, the moderator may have to work to translate the architect's input during block or unit-level planning sessions. However, just as in the case of the software engineers, this input is very valuable.

Because architects tend to think more in terms of the overall operation of the system, the attributes they identify for each feature will often be of a more complex and statistical nature. Keep in mind that while these attributes may be more complex and somewhat harder to implement as objectively measurable metrics, it is these same attributes that will enable us to avoid very hard-to-fix system-level bugs late in the design process.

For example, when discussing a device input interface, a verification engineer may identify an attribute such as transaction type and then measure that attribute by simply defining a functional coverage group that measures how many times each transaction type has been received by the device. A system architect on the other hand might identify an attribute having to do with available input bandwidth. They may identify the feature given in Figure 6.5 for the device.

The device must service input requests, however, it can stall the bus for processing. The device must not stall the bus more than 10 cycles out of every 100.

Figure 6.5 Architectural Feature Description

This device feature may not have been evident without the system architect present. While it will be more difficult to implement the metrics that measure the attributes of this feature, the development time spent now will save the time required to debug the device when system-level simulations discover the feature was not properly implemented.

An experienced moderator will compare and contrast the architect's understanding of a given feature to the design engineer's understanding of the same feature. In fact, in projects where verification begins at the architectural level, great benefits can be reaped by performing architectural verification planning sessions. The session is a bit of a role reversal compared to the design-centric planning session described above. In an architectural planning session, the system architect diagrams the system and is the key engineer being interviewed. The design engineers present at the planning session come away with a greater insight into how each of their blocks fits into the system as a whole.

Now that we've considered the planning process with respect to each of the contributors, let us take a look at how the process changes with regard to each level of system integration. But before that, let's take a quick look at reuse opportunities as the level of integration changes during the course of the project.

Reuse Considerations for Planning Over Different Integration Levels

As we move from one level of integration to another during the course of the project, our coverage concerns will change. For example, we will need to verify that we have 100% coverage of each available transaction type on a device's input bus during block-level verification. However, as we move to unit and chip level, this consideration which was verified at the block level will have less importance. At the unit level, we may be more concerned that the interfaces between our block and the blocks adjacent to it have all been exercised in their most common customer usage modes.

By taking these considerations into account at the planning phase, we can construct more effective plans and utilize our verification engines more efficiently. As we identified above, not every feature attribute carries the same weight over the entire course of the project. By applying uniquely specifiable goals to each feature attribute for each phase of design integration we can keep the plan in better perspective. Using the example above, we might attach a coverage goal to the transaction type of 100% at the block level of verification. However, we might specify a goal of only 30% integration verification. We know that all transaction types have been verified at the block level. At the integration level, this metric may serve only as a second check that the device is being sufficiently exercised.

In this case, our sample verification plan would look like Figure 6.6.

In the above verification plan, we have specified two different views into the available metric data. The views will display the annotated coverage information described in the section of the verification plan that they reference. The views will grade that data based on a separate goal defined by the "'goal" directive. In this manner, we will see different completion grades for the project based on what phase of the project we are currently executing.

```
Device Input Interface
    Burst Transactions
The device can accept burst transactions in
sizes of 2, 4, and 8 words. When the burst_n
signal transitions to high, a burst transac-
tion has been completed. By sampling the bsize
bus, the burst size can be determined.

cover: /sys/interfaces/dma_in/burst_trans

View: Block Level Verification
reference:  Device Input Interface/Burst Tran-
sactions
goal: 100%

View: Unit Level Verification
reference:  Device Input Interface/Burst Tran-
sactions
goal: 30%
```

Figure 6.6 Verification Plan Excerpt with Views

Finally, it should be recognized that some metrics will be of no value during certain phases of integration. For example, transaction type coverage may be of little or no value during system integration when the key focus is on testing the integration of the device and its associated firmware and application software. Likewise, coverage of the firmware routines that have been executed is probably meaningless at the block level of integration.

With this in mind, care should be taken to architect verification environments so that coverage metrics can be added or removed at the appropriate levels of integration. This ability will serve both to reduce the complexity of the verification environment and to improve the performance of the various verification engines used. This recommendation can be realized in several different ways dependent on the verification tools and languages being used.

Now, let's look at how the focus of the planning process changes for each level of design integration.

Planning for Block-Level Design
Block-level verification uses the finest level of detail. At this level we have the most observability and controllability of the design. It is here that we can most easily verify the smallest details of product intent. Consequently it is at the block level the most white-box features will be revealed. Be careful to dig for them if necessary. Don't forget to query all stakeholders about any corner cases of functionality that concern them.

The moderator should remember not only to focus on what the block under verification will do (what features it implements), but also to focus on how those features will be used. The tendency may be to focus mostly on the design and verification engineers at this level of planning. Be certain to involve the architects and software engineers as well because of the uniquely valuable perspectives they offer.

Keep in mind the verification tools that can be brought to bear on this phase of the project and the contributors that drive these tools. The tools typically available for block-level verification are:

- Assertion-based formal verification
- Simulation-based dynamic verification and testbenches
- Assertion-based dynamic verification
- Reusable libraries of verification IP for interface protocols
- Reusable libraries of verification IP for device features
- Accelerators
- Emulators

Accelerators and emulators are less-frequently used and are included here for completeness.

As the planning session proceeds, query the originator of each attribute or verification concern as to how their concern might be most easily measured. Keep in mind that we want to have as much of the team working in parallel as possible. Look for opportunities to start design engineers on verification by using formal tools before a testbench is even available. Look for concerns that might already have metrics implemented in a reusable verification IP library. At a planning session, you have the collective mind behind the project assembled in one room. Don't forget to query them all as to how a metric might be most easily measured. But, once again, be careful to never let a verification planning session evolve into a discussion of *how* to verify a feature.

Planning for Integration Verification
At this level of verification, we are testing an assemblage of blocks for correct behavior. There are two major concerns here. First, do the blocks communicate with each other in an appropriate and correct manner? Second, does the subsystem of blocks correctly implement the features as they were intended?

When integration planning, it is important to have the designers of each of the blocks in attendance. The facilitator should question the designers in a way that ensures they have the same understanding of the interfaces between their blocks.

At this level, it is also crucial that the system architects and the software engineers are in attendance. This will be our first opportunity to verify that the subsystems operate as intended by these two groups. They can add insight into the intended operation of the subsystem that we can't get from the design or verification engineers.

The planning session's emphasis should be on the features of the integration as a whole. The features of each individual block should be considered less if at all at this level. By using the status reported by our block-level verification plan we can show that these features have been verified at the block level of integration. This focus will make implementation more efficient and will also improve the efficiency of our verification engines because we are using them only to verify the current important focus rather than taking the entire project into account.

Planning for Chip-Level Integration
Chip-level planning focuses on communication between the different subsystems, and the features that are implemented by the chip as a whole. Chip-level planning should focus on the system architects and the software engineers. These are the stakeholders that can tell us the highest priorities and device configuration to be verified. With the explosion of possible usage combinations at the chip level it becomes essential to prioritize our activities in this manner.

The moderator should try to reveal corner cases that are exacerbated by different design subsystems acting concurrently. These concurrency cases should be tracked using functional coverage. When constrained random testing is used, appropriate coverage is of the utmost importance. With functional coverage metrics the team can detect corner cases that have been exercised without having to write specific directed testcases to target them.

Planning for System-Level Integration
The focus of system-level integration planning should be on the interaction between the device, the firmware and the application software that run on the device, and the other devices that will either drive or be driven by the device. Once again we focus on the system architects and the software engineers. In addition to these stakeholders, we may also include marketing representatives in the session because of the unique perspective they have on how customers will make use of the device in their systems.

Feature attributes are most often described in terms of the applications that will most typically be executed with the device. The metrics are measured in terms of software coverage, and assertion coverage.

We've looked at what each stakeholder can contribute and how the plan changes as the stage of integration changes. At this point, we have a plan almost ready to be executed on. But first, let's make it a more meaningful plan by attaching goals, milestones, and views.

Decorating the Plan: Views, Goals, and Milestones

Now we know what we want to verify and how to measure that the job is complete. But, what are we going to measure our progress against? It's time to add goals, milestones, and views to the plan.

Each stakeholder will have different concerns regarding our verification projects and should be able to view the plan's metrics based on their concerns. This variety of views is shown in Figure 6.7. In order to keep all stakeholders involved with the project, we'll want to enable each of them to view the data that is important to them in the format that they desire.

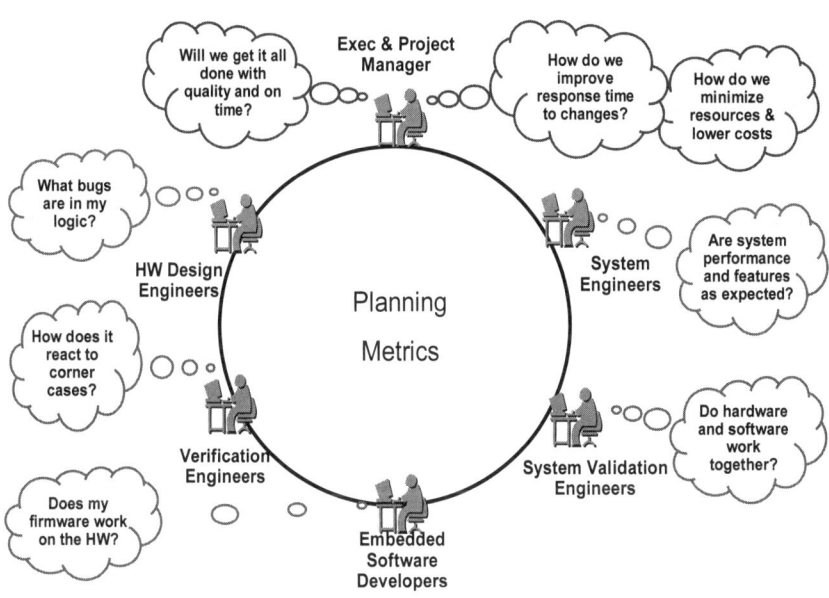

Figure 6.7 Stakeholder Views

Remember, we can look at the plan from multiple perspectives, or views. After the planning is complete, we need to define a view for each stakeholder so they can easily track their concerns and provide feedback as necessary. To define a view, we just declare the view in the plan document and then add references to the metrics that we want to be included in a view. The plan snippet in Figure 6.8 illustrates this.

View: Designer Verification
reference: "Device Input Interface/CodeCoverage"
goal: 93%

Figure 6.8 Verification Plan View with a Goal

A view called "Designer Verification" has been defined. The only metric data that will be displayed in this view is the code coverage results for the "Device Input Interface" block. Similar views can be defined for each stakeholder.

We have placed a goal of 93% coverage on this metric. We have a goal, but when does the goal need to be completed by? We can add a completion milestone to the plan with another annotation shown in Figure 6.9.

View: Designer Verification
reference: sys/Device Input Interface/CodeCoverage
goal: 93%
milestone: 4/06/2008

Figure 6.9 Verification Plan View With a Milestone

Now with a goal and a milestone, we can track our progress over the life of the project. We can chart graphs like the one shown in Figure 6.10.

Here we see the status of code coverage collected from every weekly regression. The goal for code coverage is shown by the horizontal line at 93%. The deadline or milestone is shown by the vertical line at 4/06/08.

6 Verification Planning

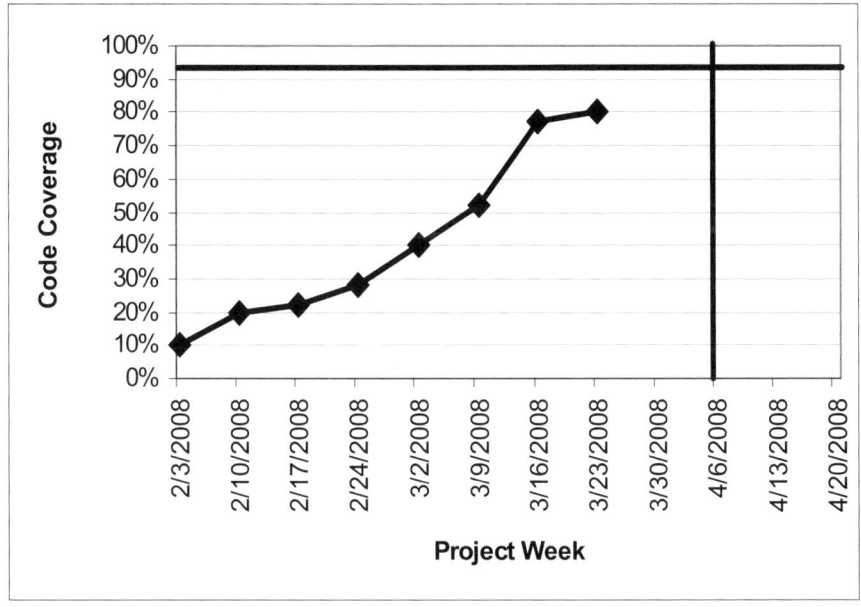

Figure 6.10 Tracking Design Completion

Summary

In this chapter we learned how to perform metric-driven verification planning. We walked through the fundamentals of the brainstorming session that provides the raw data for our executable verification plan. Then, we looked at how the brainstorming process should be modified keeping each stakeholder and each level of project integration in mind. Finally, we learned how to partition metrics into views and attach goals and milestones to them.

Chapter 7
Capturing Metrics

Introduction

Excited about the ability to use metrics to automate your verification processes? Tired of waiting on a standard to arrive so you can organize all your coverage metrics in one place? A methodology is presented for aggregating all your metrics *now* so your verification projects can benefit *today*!

Metric process automation (MPA) tools are offering huge gains in the productivity, predictability, and quality of today's verification projects. These tools automatically collect the metrics that are generated by your verification engines such as simulators, emulators, and accelerators. Using these metrics you can track the status of your project and automate processes such as debug triage, coverage ranking, and status reporting (Figure 7.1).

Using a top-down verification planning technique, you can determine what metrics each team member associates with each feature of the device. Using this customizable metric selection, each user can track the project based on their concerns. Each team member cares about different metrics lots of metrics. Let's take a look at some of those metrics. How about just the coverage metrics:

- Code coverage
- Functional coverage
- Assertion coverage

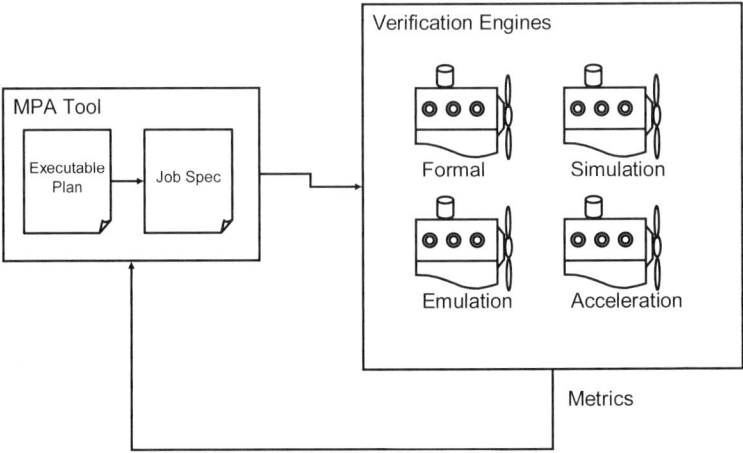

Figure 7.1 Automating Verification with Metrics

Each of these metrics can come from a different tool. Each of these tools could come from a different vendor. How do you merge all this data? Will the different formats from different vendors match? This is exactly what's being addressed by the universal coverage database standardization effort. There are other questions though. If a vendor has already optimized an analysis engine for their coverage data, shouldn't you be able to take advantage of those optimizations? What if a new kind of coverage emerges that nobody thought of?

OK, for now, let's hope the standard will straighten some things out (at least for coverage metrics). But you have a project to complete today. Wouldn't it be nice to use the most advanced techniques right now? No matter what metrics you choose? What if you could preserve the optimized performance of the tools you're running right now without modification? This chapter presents a methodology that will let you do just that.

The Universal Metrics Methodology

Using our MPA tools, we'd like to:

- Automatically collect metrics from a variety of tools.
- Display those metrics in a manner that is meaningful to each stakeholder.
- Use those metrics to automate appropriate processes.

The value of a metric process automation tool is in the collection, and utilization of metrics. Where the metrics come from should be transparent. Analysis of the metrics can be handled by the tool. However, the tool should be able to take advantage of existing analysis engines as well. There's no need for the MPA tool to take over all the analysis tasks.

The idea of using multiple applications to handle tasks has been around for awhile. Consider the way a web browser handles image and animation data. There are at least two ways the browser can display the image. First, the browser could read all the bits of the image and store them internally. Accessing the image bit by bit, the browser developers could implement image display code. The browser implementers probably wouldn't want to create a new image display module for each new image format. Perhaps they could force all image providers to conform to a single image format standard. And when a more compact image format comes along that allows faster surfing? Well, let the standards board worry about that!

There's an easier way though. By implementing a well-defined interface, the browser can defer the image display to another application. The browser simply forwards the image bits and a reference to an available screen area to an image display application. That application interprets the data and displays the image in the space allotted. Using this methodology, the web browser's implementers don't have to implement new image-processing functionality. They can benefit from the effort of others through application reuse. And, that's exactly what we'll do with our MPA tool.

First, we'll define the set of operations that we'd like to perform on coverage data:

- Display coverage data as a completion percentage along with the size of the coverage space and the number of hits within that space.
- Correlate coverage data back to its source. In the case of simulation for example, we'd like to correlate coverage data back to the testcase that created it.
- Rank coverage data against its source so that we can find the most efficient sources for creating unique coverage data. Using this information, we can create efficient regression suites.

Now, we'll define a standard interface for our MPA tool to talk to other applications that can create metrics. Once we have this standard interface in place, we can immediately start to benefit from our efforts. Based on our list of operations above, our interface might look like Table 7.1.

Table 7.1 Metric Linking Interface

Method	Explanation
list of coveritems get_cover_items()	Returns a list of structures that describe coverage items. Each cover item struct consists of a name, and a set of dimensions
get_coveritem(name: string)	Returns the number of hits for an individual item
list of source correlatecoveritem(name: string)	Returns a list of source structs that describe the simulation that produced the coveritems
list of source rankcoveritem(name: string)	Returns a list of sources sorted by the number of unique buckets hit within the coveritem's coverage space

7 Capturing Metrics

Now, as new metrics are defined that come from new sources, we simply have to provide a small interface of functions that translate the metric data from existing applications into a format that our MPA tool can understand. In object-oriented programming circles, this is known as the proxy design pattern. This proxy interface is easy to implement. It's much easier than modifying our MPA tools every time a new metric is required. And, it's certainly much easier than asking all our existing vendors to update their applications to conform to a standardized data format. In fact, any existing tool that produces ASCII-formatted output reports can be easily adapted to this interface by an end-user with reasonably proficient programming skills. Figure 7.2 shows a graphical representation of this methodology.

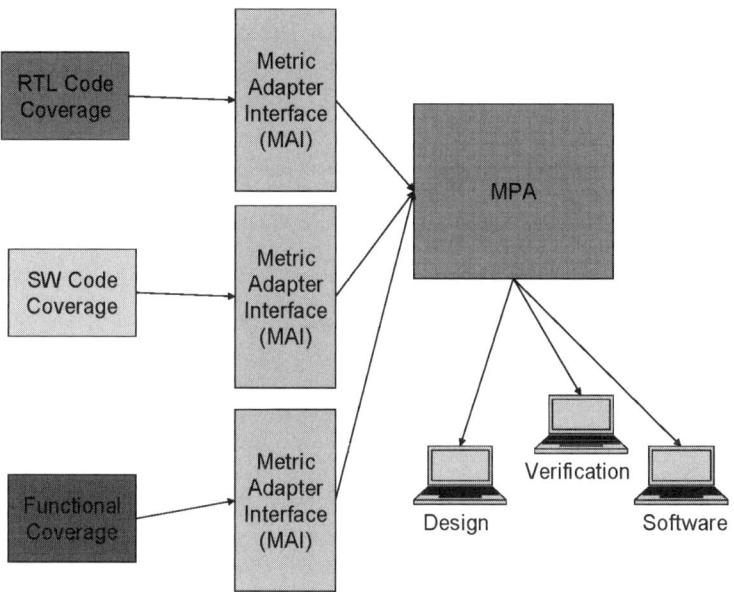

Figure 7.2 The Metric Adapter Interface

This is a simple concept that's been used for years in office automation software. Using this concept we've shown how to reap the benefits of the optimized analysis engines you've already purchased in combination with the latest MPA tools available today.

Chapter 8
Regression Management

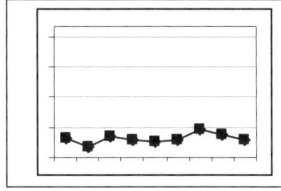 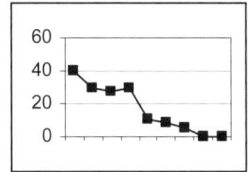 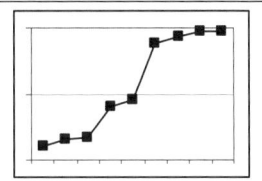

Introduction

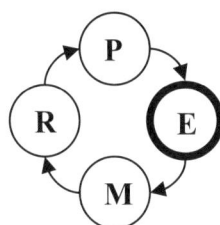

Our definition of regression is any execution of verification processes that certifies the quality level of a portion of the design. Regression is the key mechanism for gathering the objective metrics that drive all our verification processes. Regression management is the task of launching verification jobs using the appropriate metrics and tracking the metrics that are returned from those jobs. These metrics can then be processed and analyzed to facilitate a metric-driven process automation flow.

Regression management is intertwined with revision management. We discuss revision management in depth in the following chapter. Simply put, revision management is the process of managing released design code to ensure that the project team can back up to earlier revisions of the code and to ensure that code that meets certain quality criteria can be easily accessed.

In this work, regressions are divided into the two rather broad categories of revision management regressions and design quality regressions. Revision management regression as discussed below are used to check the design's base level of functionality before releasing it to the general engineering community. Design quality regressions are intended to exhaustively qualify the quality of the design for production release.

Early Regression Management Tasks

You should run regressions at the same times your local crooked politician would like you to vote: *Early and Often!* Regression management is crucial. This is where the majority of metrics that will be used to adjust the verification process will be gathered. Regressions should be started as early as possible. There are two advanctages to this. First, we gain visibility into the progress of the project via measurable metrics at an early stage. This visibility makes stakeholders more engaged in the project on a day to day basis.

The second advantage is more pragmatic and applies to projects just beginning to use metric-driven processes. By starting early the regression apparatus itself is tested at an early stage of the project when its proper operation is not crucial. The project team can detect and fix problems in the regression system before they impact more crucial stages of the project.

Among the metrics that should be tracked almost immediately are:

- Number of lines of design code checked into the revision control system.
- Number of changes made to released design code.
- Number of lines of verification environment code checked into the revision control system.
- Code coverage of the simulated design.
- Functional coverage.
- Formal assertion-based coverage.

Regression Management

Regression management is crucial to the success of the project. Regression technology is rather benign and it's not rocket science. However, as with most simple aspects of a project, if regression management is ignored or mishandled, it can consume man-weeks of project time. Several regression strategies have been outlined in other sources. Several tool vendors sell regression management solutions. They all work relatively well. It is very important to standardize on a regression management framework as early as possible in the project and then stick with it. Regression management can become a religious issue quickly. All engineers have worked with systems they did or did not like. Everyone has an opinion. However, the true value of your design and verification team is the *stellar work they can do on your chip design*, not the value they can add by picking the perfect regression management system. Listen to their input and weigh its value once. Then, pick a system and get busy with the important work of designing the chip.

The regression management framework we show here is simplified to demonstrate what metrics should be tracked in regard to "typical" regression and revision management activities. The true core value of the content is intended to be the proper application of metrics to the regression process. The infrastructure presented here is not meant to be taken as an authoritative representation of the "best" system, it is merely provided as a demonstration vehicle.

Linking the Regression and Revision Management Systems

Regression management is tightly linked with and supported by revision management. Figure 8.1 shows the layers of revision management used for the linked revision/regression management strategy discussed here.

The engineers develop code in their own insulated personal revision areas. Within these areas users are free to experiment with different version of source files as necessary. When engineers believe their code is ready for integration, they run a bring-up regression to qualify their code for release into the integration revision area. In the integration revision area, the engineer's code is further qualified to ensure that it "plays well with the other children" and can be released to the general population. After a set of integration regressions, the code is promoted to the released revision area where it is available to other users and can be tested for production readiness. Code is promoted from area to area with the use of associated revision management regressions that are discussed below.

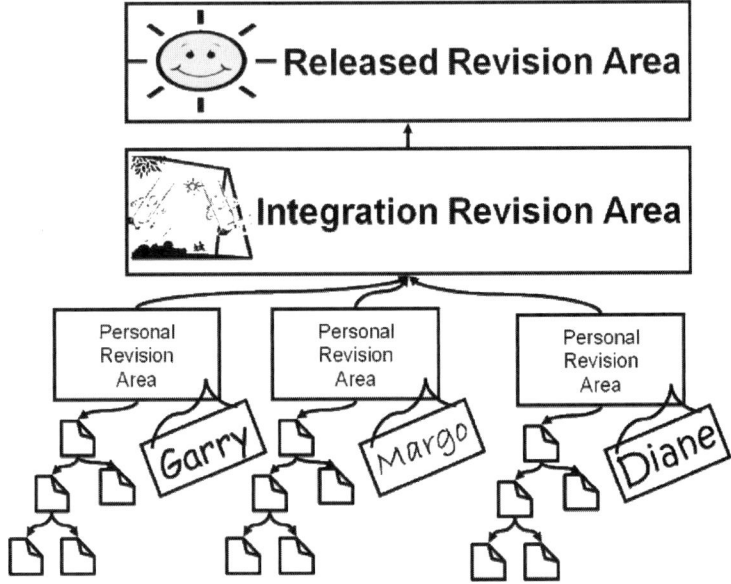

Figure 8.1 Revision Control Areas

Bring-Up Regressions

Bring-up regressions are used as a base-line to certify that new design modules are basically functional before releasing them to the rest of the project. Some aspects that should be checked in a bring-up regression are:

- Does the code compile and simulate?
- Is the basic interface functionality of the block alive?
- Do all formal assertions defined for the block still pass?

These regressions should be run every time the module is modified and before the new module code is checked into the integration revision control area. Because they may have to be run very frequently, these regressions should, of necessity, be very brief in nature.

Bring-up regressions should be run in the developer's personal revision area. In this manner, the rest of the project is insulated from changes made to the module until the module passes bring-up regression and is promoted to the integration revision area.

8 Regression Management

Metrics should be tracked for these regressions. Some of the key metrics are shown in Table 8.1 along with the trends that should be observed for these metrics over time.

Table 8.1 Bring-Up Regression Metrics and Trends

Percentage of code exercised, (simple code coverage)	*graph ranging 70–100*
Length of regression	*graph, relatively flat low line*
Functional coverage	**100%**
Number of nonsynthesizable warnings per regression	*graph decreasing from ~40 to 0 (range 0–60)*

Code Coverage

Code coverage is a simple metric that can be used to gauge bring-up regression completeness. Its purpose is merely to ensure that the test suite continues to check the entire block as new features are added. Code coverage doesn't need to be 100% at this stage, but should wiggle around a relatively high percentage.

Regression Length

As mentioned, the length of a bring-up regression should be relatively short. If it runs too long, its usefulness is reduced and may ultimately be eliminated if engineers opt not to run the regression at all in favor of accelerating the release process by simply checking their code in untested. The regression length should remain short over the duration of the project. This metric should be checked with each regression. As blocks grow in feature count, the simulator can slow down, noticeably increasing the total regression time.

Functional Coverage
With the use of a verification plan perspective or view, the functional coverage of these regressions should always be 100%. There is a certain base level of functionality that must be guaranteed before a block should be released to the general population. A sample verification plan for a bring-up regression is shown in Figure 8.2.

DMA Block Reset Behavior

```
All signals should be set to their reset
value and all FIFOs should be flushed.
cover: /sys/dma/reset_beh/*
```

DMA Registers

```
All registers should be readable and
writable. All masked bits (write only, read only,
etc.) should behave appropriately.
cover: /sys/dma/register/*
```

DMA Configurations

Top Customers
```
DMA transactions should work in the top three
customer configurations of the DMA engine.
cover: /sys/dma/top_cust_conf/traffic/*
```
Others
```
All other configurations should be exercised.
cover: sys/dma/other_conf/traffic/*
```

View: DMA Bring-Up Regression
```
reference: DMA Block Reset Behavior
reference: DMA Registers
reference: DMA Configurations/Top
           Customers
```

Figure 8.2 Bring-Up Regression Verification Plan View

The verification plan above shows the use of a view to only reference the portions of the plan that are important for bring-up regression qualification. For example notice that we have restricted our bring-up view to only show the "Top Customer" configuration. The functional coverage for this view should be 100% on each bring-up regression.

Nonsynthesizable Warnings per Regression
Early in the design cycle some behavioral constructs are acceptable to accelerate the modeling process. However, as the design approaches production release near the end of the project these constructs should be eliminated. By automatically measuring warning messages about behavioral nonsynthesizable constructs provided by linting tools, the team can verify that they have been removed.

Integration Regressions

Integration regressions should take place in the integration staging area of your revision control system. These regressions are crucial. They ensure that each module of the design will play nicely with all the other modules and *at the very least* not break the released simulation and regression flow. Countless man weeks of progress have been lost on many projects because this simple regression step was skipped by developers eager to get their "fixed" module back into the regression flow.

Table 8.2 Integration Regression Metrics and Trends

Functional coverage of available transaction types between connected blocks.	**100%**
Functional coverage of scenarios that are executed by a combination of blocks. Especially where one block makes use of specific data provided by another in performing its function.	**100%**
Length of regression	

The integration regression asks the question "Does this block integrate well enough to not break the integrated design?" Less emphasis is placed on the individual functionality of the block (presumably that was tested in the bring-up regression), and more emphasis is placed on the interactions between the block and other blocks in the design. Functional coverage is more meaningful and easier to interpret with respect to these interactions than code coverage. However, code coverage can still give us a guarantee that all blocks involved were indeed activated. Some of the metrics that should be tracked for this type of regression are given in Table 8.2.

Functional Coverage of Bus Traffic
One of the key questions during integration is, "Do the blocks talk?" This can be easily answered at a basic level by using functional coverage to track the types of bus cycles that are transmitted between blocks during the integration regression. It is important to capture *every* type of bus transaction that can be propagated between blocks in the executable verification plan. Also keep in mind that blocks do not have to actually share signals to be in communication with each other. One of the important cases in which traffic needs to be tracked is the case where many masters can communicate with many slaves over an arbitrated, address decoded bus as shown in Figure 8.3.

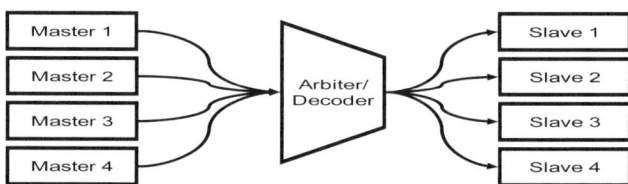

Figure 8.3 Masters Slave Cross Traffic

Integrated Block Scenarios

Most of the block-level functionality should be tested elsewhere. However, it is important to check interblock scenarios in these regressions. Two of the inter-block scenarios that should be considered are:

- Scenarios where one block is dependent on accurate, correctly formatted data from another block to perform its function.
- Scenarios that stress bus bandwidth where blocks are dependent on a guaranteed flow of input data.

Length of Regressions

Here again, the regression team is forced to walk the fine line between testing enough functionality to guarantee successful integrations, and keeping the regression short enough that it will be effective. By historically tracking the length of the regression, teams can modify the regression suite as necessary as total regression time increases.

Design Quality Regressions

Design quality regressions are intended to exhaustively verify released functionality. In design quality regressions, the goal is to fully populate the executable verification plan with complete coverage of all identified stakeholder concerns.

There are a number of coverage metrics that should be measured with respect to this type of regression. These metrics will be described in much greater detail along with the various technologies they are associated with in Part III of the book. They are included here as a reminder of what metrics to consider when performing verification planning and setting up the regression and metric tracking apparatus for the project (Table 8.3).

Table 8.3 Some of the Metrics and Trends for Design Quality Regressions

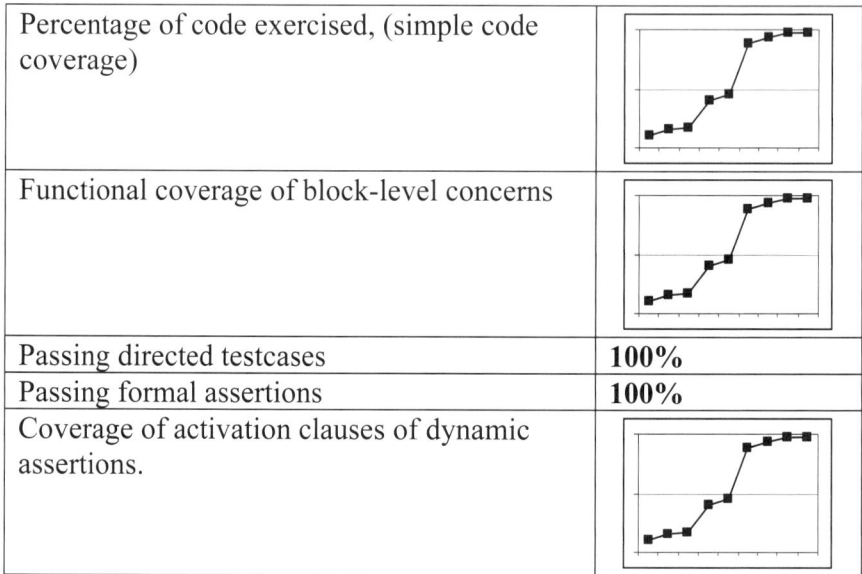

Percentage of code exercised, (simple code coverage)	
Functional coverage of block-level concerns	
Passing directed testcases	100%
Passing formal assertions	100%
Coverage of activation clauses of dynamic assertions.	

Managing Regression Resources and Engineering Effectiveness

Engineers often perform their tactical testing using the same compute resources that support regression tasks. It is important to remember to allocate sufficient compute resources to our engineers so they can efficiently perform tactical regressions such as revision promotion

Table 8.4 Metrics Used to Track Tactical Regression Effectiveness

Utilization of compute servers	
Wait time for tactical job start.	

regressions and debug regressions. The metrics that are provided by our metric-driven processes can be used to gauge this allocation. Two of the metrics that should be monitored are given in Table 8.4.

Regression managers need to make sure that compute servers are as fully utilized as possible while also ensuring that no engineer has to wait an unreasonable amount of time to start a tactical regression job. By tracking the above metrics, the regression manager can respond to balance compute server utilization. *Remember, one of the key values of metric-driven processes is to fix problems before they arise!*

Regression-Centric Metrics

So far we have discussed what metrics should be collected from our *execution* engines during a regression. There are metrics created by our *regression* engines that should be captured as well. Among these metrics are:

- Frequency of regressions run grouped by regression type
- Number of issues found per regression
- Machine utilization
- Software utilization

Frequency of Regression Runs Grouped by Regression Type
This metric can be used as a tactical indicator. As the project proceeds, the number of design quality regressions should increase and the number of revision management regressions should decrease. Sharp increases in the number of revision management regressions can indicate that the design has become more volatile. This is not necessarily a bad indicator. The key thing is to ask the right questions to determine *why* the design volatility has increased.

If the frequency of design quality regression decrease as the project proceeds, this could indicate that the regression apparatus is broken, or that significant amounts of time are being spent on other activities such as adding features, or debugging existing issues. Again, the point is to use these metrics as triggers to ask appropriate questions to understand the project status and then respond effectively.

Number of Issues Found per Regression

This is one of the oldest metrics and also one of the most abused. Historically, this metric was often used to determine when a chip development project was complete. If the number of issues detected were trending toward zero per regretssion, then the reasoning was that the device under test must be approaching full functionality. Of course, this ignored the possibility that the verification environment might simply be looking in the wrong places for bugs.

This metric should however, still be measured and does serve some useful purposes. When used in conjunction with coverage metrics, the number of issues found per revision can add to the confidence that a device has been properly verified. When used in conjunction with revision control information, the number of issues detected can point out blocks that may need to be re-engineered. Figure 8.4 shows issue tracking metrics displayed with revision metrics for the DMA block of a design under test. Using this data it can be seen that the design was plagued with issues after the large code change in the fourth revision. This data may lead the engineering team to decide to revert to revision three and start over. As with all metrics, interpretations should be made carefully and involve plenty of communications between the engineering team. The peak in issues after revision four could just as easily have been caused by added features that were not available for testing before that revision.

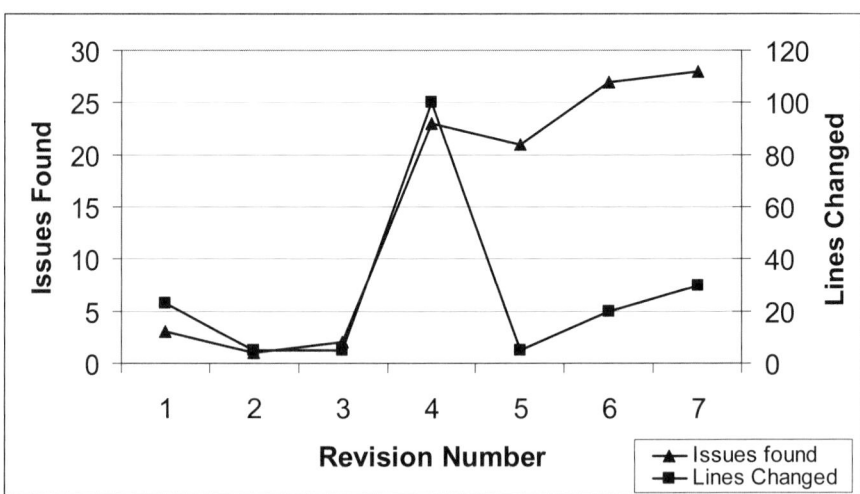

Figure 8.4 Issues Found vs. Lines Changed

Machine Utilization

This simple metric gives an indication of the utilization of compute resources on the project. Ideally, this metric should oscillate near 100%, indicating that your compute resources are being applied to their fullest value. This metric should be tempered with the wait time for an individual user to start a compute job as mentioned above. Both of these metrics can be extracted from commonly used job distribution systems.

Software Utilization

This metric is similar in nature to machine utilization, but tracks the usage of your verification software. The intent is to maximize the return on software investments. Even more importantly, this is an indirect indicator of the utilization and productivity of your engineering resources. If productivity enhancing software isn't being utilized, it may be an indicator that the engineering team does not know how to utilize the full value of the software and therefore may need training. It could also indicate that the software is too hard to use or of little value. *The key ingredient that must be mixed with objective metrics is continuous communication.* Therein lays the key to discovering the root causes of observed metric trends.

How Many Metrics Are Too Many?

Historically, when running regressions, one of the key metrics that was fretted over was simulation speed. Prior to metric-driven verification, speed was king. Because some metric collection is done during simulation, it can slow down the cycle per second performance of the simulator. This consequence led many early design teams to delay metric collection till near the end of the project. This late collection led to several interesting surprises as design teams only a few weeks away from the scheduled completion date of their project discovered that half of their design had never been verified!

There is certainly a trade off between the metrics collected and simulation speed. If too many metrics are collected and not used, the value of the verification environment can be reduced. However, the opposite is also true. If no metrics are collected and simulations are blazingly fast, we may be accomplishing nothing quickly!

There are new verification techniques such as constrained random testing that can significantly increase productivity through the effective use of metrics. In the case of constrained random test generation, the price of metric collection is easily offset by the increase in productivity that is gained by not writing testcases. This subject is covered in much more depth in a later chapter.

We can actually use metrics to determine how much our metric collection is costing us. First, by measuring the CPU time used per simulation, we have a simulator/verification environment performance metric. We can then measure simulation performance information with metric collection turned on and off. This data will contribute to our decision process of how often to collect metrics.

Objective results regarding simulation speed with and without metric collection should be considered in the context of verification closure speed. If verification on the project is completing within the allotted schedule, then there may be no need to optimize simulator speed. However, if verification is lagging behind, and there are true advantages to be gained by simulating more quickly, then these measurements should be considered.

Once it has been determined that there are valuable gains available by turning metric collection off, we can develop a strategy for effective metric collection. Keep in mind, for example, that it doesn't make sense to measure all metrics all the time. If a given block has been declared to have satisfactory closure on verification and the block has not been further modified, then metric collection for that block can be turned off.

Some reasons to turn the block-level coverage back on are:

- Any modification of the RTL for the block no matter how trivial.

- Modification of blocks that communicate with the block whose coverage has been turned off.
- Addition of new coverage groups into the block based on new or modified features of the device.
- Detection of new failures within the block.

Summary

In this chapter we have reviewed the various types of regressions, their importance and some of the metrics that should be tracked during regression activities. We have also outlined the basic relationship between regression management and revision control activities.

In Chap. 9 we will look at revision management and the metrics it produces and consumes.

Chapter 9
Revision Control and Change Integration

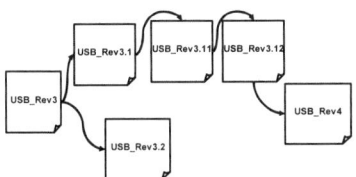

Introduction

When discussing revision control systems there are a few key terms that will be frequently used. These terms are defined here to avoid ambiguity.

Revision Control System: A revision control system is any application that allows the user to save multiple versions of a file so that older version of the file can be easily retrieved. Examples of revision control systems include full-blown systems like CVS (which interestingly enough is freeware), and the revision control system built into Microsoft Word that allows authors to monitor changes made to a document (which was used in the creation of this book).

Revision: A version of a single file. The user creates a specific version of a file by "checking it in" to a revision control system. Each revision is a unique version of the file that corresponds to a development step in the project. Revisions are used to allow developers to access design files when they were in an earlier or alternative stage of the development. Revision control systems allow the user to insert a brief message that describes the version of the file (Figure 9.1).

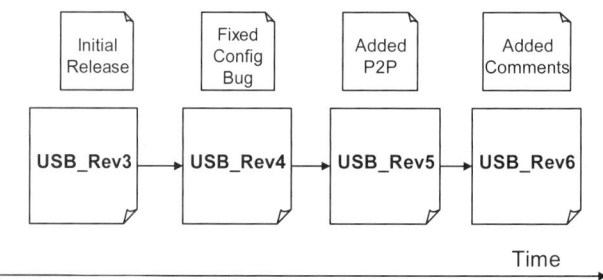

Figure 9.1 File Revisions

Branch: Frequently it is necessary to create multiple versions of a development file at the same level of the development process. One example of this is for functional debug. A developer may have several independent code changes that need to be tested to isolate the cause of a functional bug. The developer can edit the code independently for each change and then check each change into a different branch of the same revision of the file (Figure 9.2).

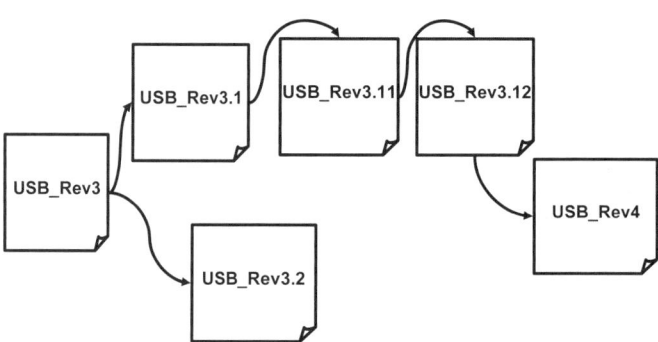

Figure 9.2 Revision Branching

Revision Tag: A revision tag is a label that is applied to a given set of file versions by the revision control system. The tag is used to link sets of files together at a common development point. One example of using a tag is to label a set of files that have successfully passed an integration regression (Figure 9.3).

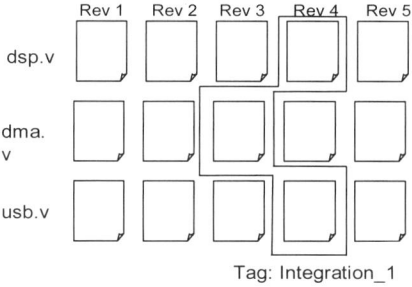

Figure 9.3 Revision Tagging

The Benefits of Revision Control

Revision control provides several pragmatic benefits. Two of the key benefits are:

- *Retrieve earlier revisions of the design or verification environment*

This is crucial for debug activities. Being able to easily try the same testcase with earlier versions of the design can provide key debug information.

- *Manage multiple versions of the same code for debug and development experiments*

Today's large compute farms allow some innovative debug and development opportunities. Rather than trying one debug or development experiment at a time in simulation, developers can try multiple plausible experiments at once. The revision control system can be used to create multiple "branches" of a given revision. One experiment can be tested on each branch. Once the decision is made about which branch will be used to proceed, the branches are collapsed into the next released version of the file. The revision control system allows the user to revert back to any of the experimental files if necessary.

Metric-Driven Revision Control

Revision control tools were among the first metric-driven process automation tools. Using these tools, an engineer can automatically revert back to any execution stage of the design and eliminate unprofitable changes or perform experiments.

While engineering teams have been using these tools to various degrees for years, they have rarely received the full value that these tools offer. To reap this value, engineering teams must treat revision control just like any other metric-driven process. They must first plan, and then execute on those plans.

Planning for Revision Control: It's Not Just for Source Code Anymore

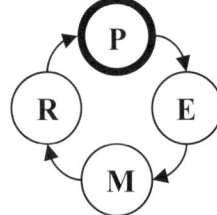

Planning for revision control means determining what will be revision controlled. Some of the planning steps are simple. For example, source files are typically revision controlled while the resulting binary executables created from the source files are not. Other aspects of revision planning are not as obvious.

One of the goals of revision control is to be able to return to a previous stage of the project and recreate the target code and environments exactly as they existed at that stage. While the user can easily recreate the same source code, recreating the same execution environment is a different matter. Changing tool versions and other metrics that aren't contained in the source code can throw a wrench into the works. But, effective planning can pull it right back out. As we said, the source code itself is certainly the necessary basic material that must be revision controlled, but teams should also revision control an *environment definition file* along with the source code. This environment definition file should contain information such as:

- Version numbers of the tools used. Tools include such things as simulators, C++ compilers and linkers, add-on verification engines, etc.
- For random testing, the random seeds that created a given constrained random verification environment
- Operating system version
- And so on

Using this environment definition file we can recreate the exact environment that was in place when the revision was stored. When planning the environment definition, it is important to consider all aspects of both the tools and the source materials that compose a design verification environment.

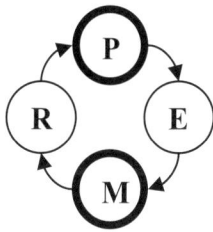

The planning done to define the environment's definition is identical to the planning required for the "process metric package" described in Chapter 10. Keep in mind that while the team needs to define what should be recorded in the environment definition file, it is not necessary or even beneficial for them to record that information manually.

MPA tools can extract and record some if not all of this information for us automatically. For instance, all of the metrics listed in the bulleted list above can be automatically extracted by MPA tools. The benefit of this is twofold, first, we don't have to spend our own time recording these metrics. This is a huge contributor to the success of these processes. Any time a person has to streamline their operations, record keeping is usually the first task to be ignored. By letting the MPA tool extract these metrics, we remove it from the engineer's list of tasks permanently. The second benefit is that we always document the exact environment that the simulation was run in. We don't document the environment that we thought we ran in, or the one we intended to run in. The MPA tool extracts these values directly from the output of the tools used to execute the verification task.

Revision Control and Documentation

Future users of the revision control system also need to be able to easily comprehend the intent of the project team, and that means they need documentation. There are several documents that need to be revision controlled on any project. Among them are:

- Marketing requirements
- Device specification
- Architectural study specifications
- Hypothetical debug spreadsheets as described in Chapter 10
- Individual design and verification notes
- Many, many others depending on the project team

Revision controlling these documents performs two functions. First, it protects the documents for posterity. No more looking around in old cast-out directories for a hint of what was going on during the project. Second, revision control also preserves a temporal element that allows us to track how decisions were made and how the design and verification of the device evolved. This information can provide valuable clues for refactoring our design during a project and for streamlining our processes in future projects.

Tagging the Revision

Tagging is the process of marking a set of revision-controlled files so that a snapshot of the project as it existed at a point in time can be easily retrieved. The project eam should plan what events should trigger tagging. Some events that should trigger tagging a snapshot are shown in Table 9.1.

This is a partial list. The user should spend time carefully considering their revision tagging plan.

Table 9.1 Tagging Triggers

Event	Material to tag
Passing a bring-up regression	All block material
An integration of blocks passes an integration regression	All integration material and all material for the component blocks
The discovery of a bug	All the material for the environment that found the bug. This should include design and verification material.
The resolution of a bug	The same material that was tagged for the discovery along with documentation created during debug.
The design passes a quality regression	The entire design and all supporting material

Reporting

Finally, the project team should determine ahead of time what reports they would like to generate from their revision control metrics. This planning should be done at the start of the project like all other planning activities. Once the planning is done, the reporting mechanisms should be put in place immediately. Like all other planning and reporting activities, this will seem trivial and low priority, however, it is of the utmost importance to setup the reporting mechanisms as early as possible. The value that revision control metrics provide can only be fully realized if the metrics can be analyzed. And the metrics will be best analyzed through the effective use of automatically generated reports.

Metric-driven process automation tools provide built-in, automated reporting capabilities. Take advantage of these to easily create the planned reports. Some reports that should be considered are:

- Volatility of the code for each block, integration and chip-level environment.
- Debug tags created per design entity per revision.

- Historic reports of engineers' changes to revision-controlled material. These can be organized with different granularities from file level to design block level.

Now that we have our plans in place for how to execute our revision control activities, let's look at a brief example that utilizes those metrics.

Visibility, Visibility, Visibility!

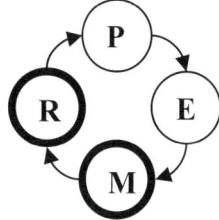

The most crucial aspect of revision control is visibility. With enough visibility into the status of a project, most revision control pitfalls can be avoided. What needs to be visible? The user should be able to see, at a glance, the status of all the files in the project, whether or not the file is checked in, or being edited, who last edited the file, and the identity of all the users who are currently editing the file.

The information that revision control tools can provide can be grouped into the following categories:

- Volatility of the code base.
- Links from code changes to engineers.
- Links between revisions and project history. Especially links between revision tags, regression success/failure data, and debug data.

Effective tracking of this information means the difference between groping in the dark with respect to what's happening with today's revision of the design, and being able to react in an effective and adept manner to changes. A veteran of even a single semiconductor design project has their share of horror stories of hours or days of regression and engineering time lost because of broken code being checked into the revision control repository.

The first sign of trouble usually comes about 8:30 in the morning when the first verification engineer arrives. The engineer, slowly coming to life with his first cup of coffee leans back and leisurely checks the results from last night's regressions. To his dismay, he discovers that every testcase the night before failed! He begins to review failure messages and discovers one of a number of failure mechanisms. A nonexhaustive collection of these might be:

- Compilation errors from a given unit ("Everything might be OK, this might be isolated to a single module")
- Elaboration or linking errors of the simulation model ("things are looking kind of dim, I might have an intermodule failure")
- The simulations all run, but there are inexplicable failures all over the chip-level regression ("it's going to take all day to figure out what caused this")

Without using revision control metrics the engineer will have to play out a tried and true routine that always follows the same steps:

- Isolate the block with the most errors and call the responsible engineer.
- "John, did you change your xyz block last night?"
  "Well, I tried a few experiments, why?"
  "Did you check it into revision control without running a bring-up regression?"
  "No, of course not!"
- Repeat.

Of course, the difficulty of this procedure will be compounded by engineers stopping by every 30 min or so to ask if the situation is rectified yet.

This situation becomes much easier if the engineer makes use of the metrics provided by the revision control systems to view the contributors that have changed various design blocks since the last successful regression. One useful display of this information is shown in Figure 9.4.

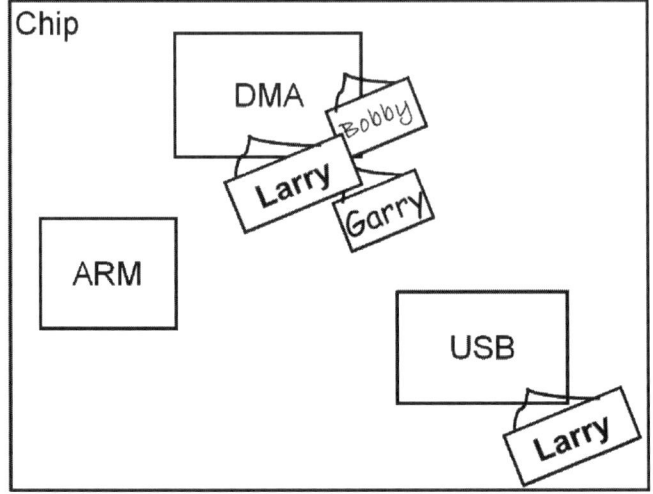

Figure 9.4 Revised Blocks

Once the regression manager has isolated a list of usual suspects, they can begin to ask more subtle questions. For example, they might pull up a timeline of revision changes including the revisions tagged as successful regression revisions. A view of this data is shown in Figure 9.5.

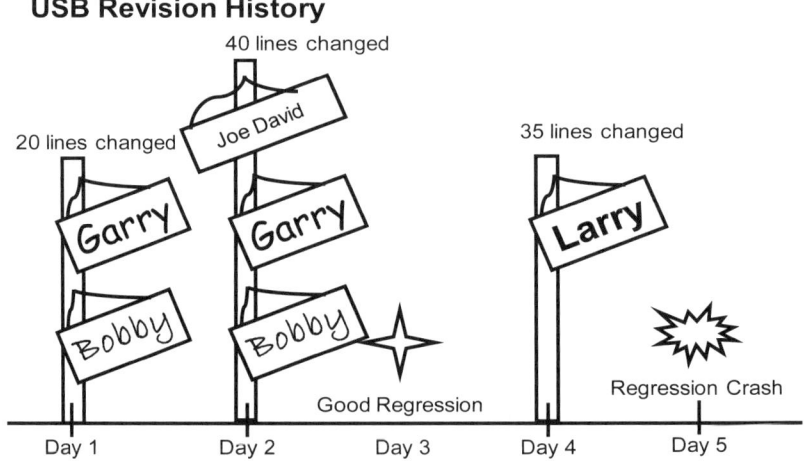

Figure 9.5 Revision History

From this data the debug engineer can see that the regression broke after the USB block was changed by Larry after the last good regression. With appropriate visibility, provided by preplanned reporting mechanisms, the debug engineer was able to quickly isolate the cause of the issue.

Summary

In this chapter, we have studied how revision control should be treated as a metric-driven process. We showed how planning for revision control as a metric-driven process ensures the best return on investment for this important activity. We also showed how effective use of our planned metric-driven reports can streamline existing processes by simply making available information *visible!*

In Chapter 10 we'll apply metric-driven techniques to debug.

Chapter 10
Debug

Introduction

Debug as a process? Absolutely! Of course, it's given, that to a large extent, debug effectiveness is based on raw talent. A hunch can lead one engineer in the right direction within minutes while another engineer might flounder in the forest of possibilities for days before bringing a bug to light. However, like any other talent-based activity (baseball, football, piano virtuoso, etc.), when talent peters out or lapses, or was never there in the first place, a return to fundamentals guarantees the most reliable road to success. And fundamentals always mean process.

Just like all good engineers, we would of course like to automate the debug process. Our objective, automatically collected metrics serve to make debug processes more efficient, and can completely automate some of them.

This chapter will describe a number of metric-driven techniques that can be used to facilitate, and in some cases completely automate, various debug tasks. While these processes offer great gains in debug efficiency, they are really only the tip of the iceberg. Many other processes can and certainly will be developed in this emerging field.

Verification Process Cloning

Engineers performing functional debug of a logic design usually engage in a process similar to that shown in Figure 10.1.

The verification engineer first categorizes and investigates detected failures to determine which failing testcases are best suited to rapid debug. They then rerun the testcase with debug information turned on to gather more information about the failure. After getting a better level of detail, the engineer studies the failure in earnest to determine if it is an actual design failure, or a failure of the surrounding verification environment. If an actual design failure is found, then the testcase is passed along with appropriate debug information to the design engineer.

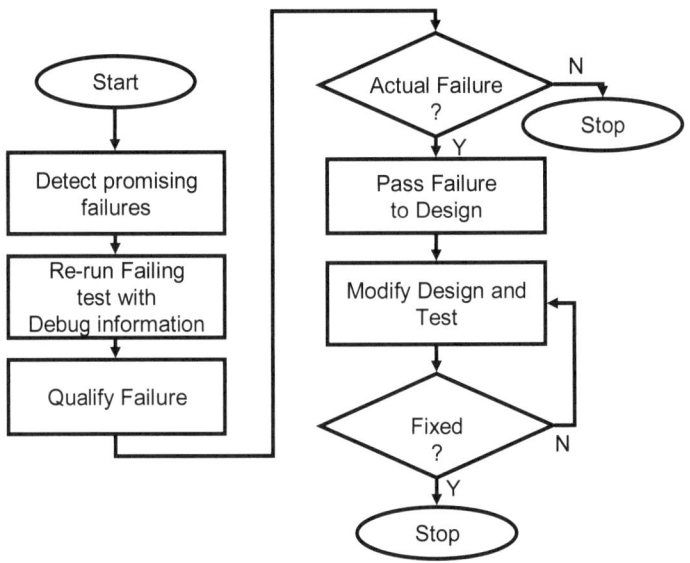

Figure 10.1 Debug Flow

This seemingly simple handoff is where a couple of hours of engineering time can be lost. Usually there's some confusion as to which version of the source code the testcase was run on. Then, there may be confusion about how environment variables were set for the simulator and the testbench. What was the LD_LIBRARY_PATH value? Where did we store the prebuilt library files for this testbench? All of this information can change from engineer to engineer.

10 Debug

All of these pieces of information also happen to be metrics. Today, these metrics are stored as what is known as cultural knowledge, or as part of what Jung called the "group mind". But when part of the group goes out for lunch, or you're not as in-tune with the culture as you could be, debug can grind to a halt.

By capturing this cultural information as a package of process metrics and storing that metric package, we guarantee that the knowledge will always be available. If we go one step further and control our processes with this metric package, the above handoff becomes completely automated. Using verification process automation tools such as Enterprise Manager from Cadence Design, we can do exactly that by metric-enabling our processes today.

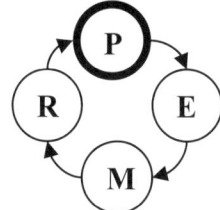

First, we plan for the process. We need to determine what metrics our process will consume and what metrics it will produce. For our purposes we're interested in the metrics that the process will consume. What metrics does our simulation engine need to successfully complete its job? A partial list of these metrics follows:

- Settings of environment variables that effect the simulation process including:
- Various path variables
- Tool-specific environment variables
- Tool command line arguments
- Tool version
- Paths to the tools that are used. Not only the simulator, but the memory modeler, the verification tool, etc.
- Revision control information. The release tag for the design and for the verification environment

This is a partial list. There may be several other metrics depending on your specific verification tool and environment setup. We use these metrics as shown in Figure 10.2.

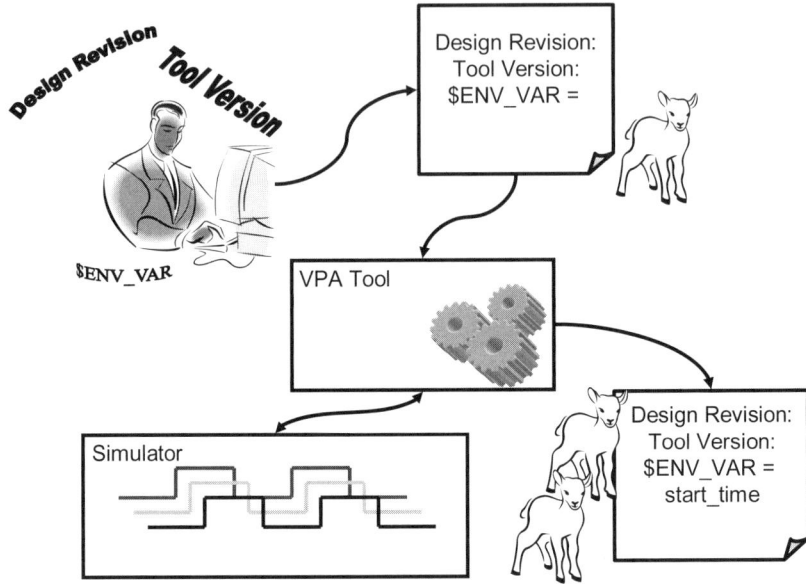

Figure 10.2 Process Cloning

The planned metrics are first encapsulated into a package that can be read by our MPA tool. The MPA tool in turn uses these metrics to drive our simulator, passing the correct command line arguments and setting the appropriate environment variables. As the simulation runs, the MPA tool creates a copy of our original metric package and adds output metrics that are specific to this simulation run.

This newly created metrics package is what we use to automate the hand-off. Using this package and an MPA tool, the design engineer can run the identical simulation on their workstation, the first time, every time. They also have all the other simulation-specific information from the original run at their fingertips, no more searching the hallways for the verification engineer.

Debug Triage
Debug triage is performed after every regression. With the size of today's design projects, regressions can consist of thousands of simulation and emulation runs. Debug triage is the process of sifting

through this data to determine the most promising failures to debug first. There are several qualifications that make a failure promising such as a failure happening early in the simulation, a failure happening as the only failure in the simulation, or a unique set of failures happening in a single simulation that don't occur together in any other simulation. To sort through the immense amounts of data produced by today's massively parallel regressions, an MPA tool is often used. The first step in analyzing the regression data is to sort the failures that were discovered. Effective error messages are crucial to this process. These error messages become yet another tracked metric. Other metrics that should be tracked per failures include:

- Failure time
- Failure module
- Failure type (assertion, check, compilation error, etc.)
- Failure description
- Testcase that created the failure
- Process metric package used to recreate the failure

These metrics can be grouped and sorted to facilitate the debug triage process.

Before debug triage is performed we can have innumerable unprocessed failure metrics. An example of raw failure metrics is shown in Table 10.1.

Table 10.1 Raw Failure Metrics

Failure Description	Type	Time	Module	Testcase
Write to full FIFO	Assertion	100	DMA	DMA_tes1
Data not written	Verification check	2000	DMA_env	DMA_test1
Write to full FIFO	Assertion	250	DMA	ALU_test33
Data not written	Verification check	3000	DMA_env	ALU_test33
Bad addition	Verification check	275	ALU	ALU_test33
Bad addition	Verification check	543	ALU	ALU_test47

By grouping on failure types and then choosing the earliest failure for each failure type, we get the subset of testcases shown in Table 10.2.

Table 10.2 Grouped First Failures

Failure Description	Type	Time	Module	Testcase
Write to full FIFO	Assertion	100	DMA	DMA_tes1
Data not written	Verification check	2000	DMA_env	DMA_test1
Bad addition	Verification check	275	ALU	ALU_test33

This view of failure metrics shows what simulations can be most efficiently rerun to perform initial debug on each failure type. The advantage of using metric-driven processes and MPA tools is that this report can be generated automatically after each regression.

Automatic Waveform Generation
For verification engineers, a chronological text trace of the transactions within a simulation is often enough for debug. However, when the verification engineer has to communicate the issue to a design engineer, there's nothing quite like good old waveforms!

Detailed dumps of waveforms are expensive in terms of simulation performance. For this reason, the first time a regression is run the waveform output is turned off. The only messages that are output are error message generated by the various checkers in the verification environment. After promising failures are found by debug triage then waveform generation should be turned back on. This generation of waveforms is often a manual "morning after" process that follows a regression. This manual procedure is shown in Figure 10.3.

In the previous section, we showed how metric-driven processes can be used to automate the failure analysis step shown in Figure 10.3. In this section, we'll go one step further and automate the entire waveform generation process.

10 Debug

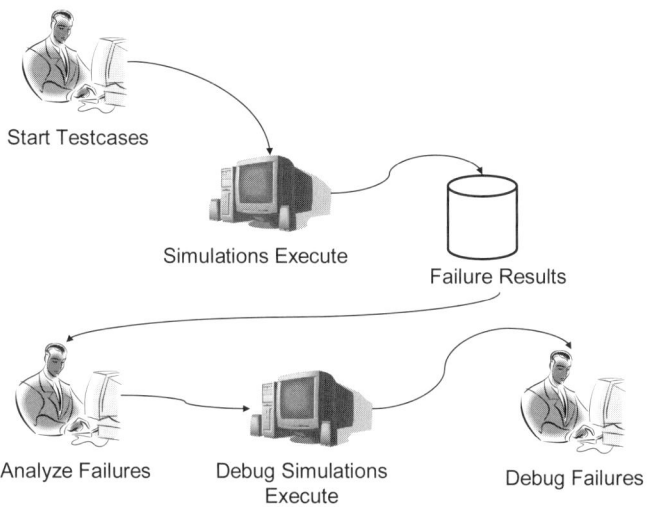

Figure 10.3 Manual Waveform Generation

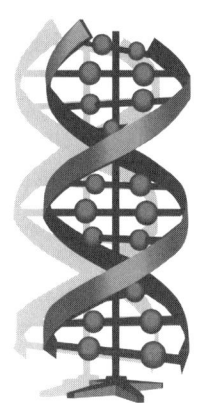

Let's dive back into genetics for just a moment. Remember that using packages of process-specific metrics we can clone processes. These cloned processes can be run anywhere and will be identical to the original process. Well, as long as we're duplicating DNA why not tweak a protein sequence or two?

When we clone the metrics package as shown in Figure 10.2, while we've got the hood open, we're free to tweak any of the metrics we like. In order to automate waveform generation, we'll first plan by defining a metric in the package called "waveform_gen." When simulations are run the first time, we'll give waveform_gen a default value of OFF to increase simulation speed and get more done with our regression resources. After debug triage, to automate waveform generation, we simply modify the process package for each simulation identified to have a waveform_gen value of ON. The process metric packages of these simulations are passed back to the execution engine and the waveform generation process is started with no intervention from our valuable engineers! The automated process is shown in Figure 10.4.

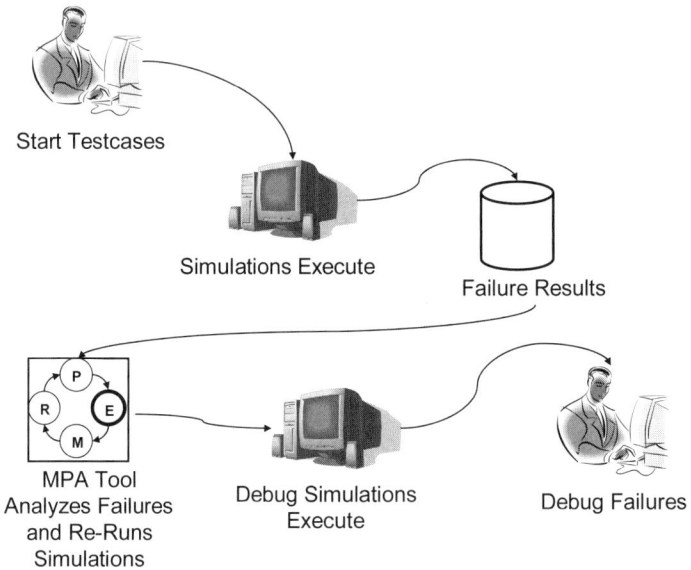

Figure 10.4 Automated Waveform Generation

Hypothetical Debug

An engineer goes to work in the morning, pulls up the automatically created trace information from a regression failure the night before, and immediately recognizes that a simple protocol rule has been violated. She fires off an e-mail to the responsible design engineer, logs the issue in the issue tracking system and moves on to the next failure. Time spent: five minutes. Great!

That's the ideal case and we all love the way it feels when we can get it. But what happens when everything goes wrong? What happens when there are multiple modules involved? What if revision control actually caused part of the problem? What if the issue is just unbearably complex? How does the engineer keep from getting helplessly lost and losing days of productivity?

Many debug sessions end in a morass of confusion and deleted files because of the tendency to try one thing after another in an undisciplined manner. Before long, it is unclear if the device behavior is truly a symptom of the original issue, or an artifact of a previous experiment. This is where a process-based metric-driven approach can save the day. The process is a hypothesis-driven debug process. This process is executed with the use of a spreadsheet, and ideally, a branching revision management system. In a complex debug situation, the engineer may have several initial hypotheses as to what has gone wrong. Some of these will naturally seem more valuable than others. The engineer captures all the available hypotheses in a miniature, individual brainstorming session. The engineer then ranks each of their hypotheses and briefly documents an experiment to prove the hypothesis correct. Each hypothesis is then proved or disproved in an iterative

Table 10.3 Example Debug Spreadsheet

Problem description:					
Data is lost from the FIFO every fiftieth frame of transmitted packets. This only occurs when the device is configured in the extended DMA mode.					
Hypothesis	Experiment	Likelihood	Results	Revision Tag	Log Pointer
The configuration register for the extended DMA mode was broken on the latest revision	Use the previous version of the configuration bridge	M	The issue is still detected	Iss179_Ex1	/regr/exp 179_ex1/ logs
The FIFO is mishandling an overflow corner case caused by the size of the transmitted packets	Check the size of the transmitted packets and modify if necessary	H	The issue goes away when the packet are made smaller	Iss179_Ex2	/regr/exp 179_cx2/ logs

process based on the initial ranking. Any new hypotheses are recorded and ranked during each iteration. Iterations are repeated until the root cause of the bug under study is determined. This process is illustrated in Table 10.3.

By using a revision tracking system, the classical experimental method of making a single change and collecting data can be more effectively implemented. By procedurally requiring that each individual experiment takes place along an independent revision branch, the confusion described above can be completely avoided. Of course, some experiments naturally lead to others and require that additional steps be taken. These can be facilitated by iterating version numbers along a revision branch (Figure 10.5).

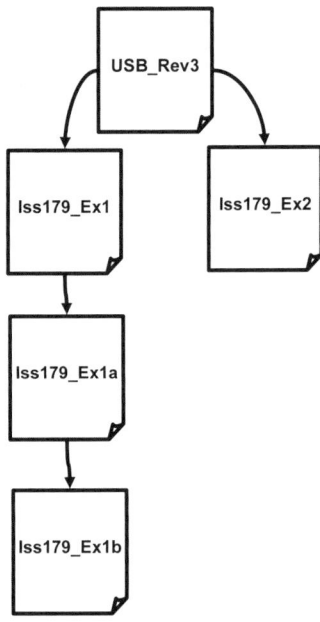

Figure 10.5 Revision Branching Hypotheses

In addition to keeping experiments along easily manageable orthogonal branches, the revision control techniques mentioned above allow a more effective use of the parallel debug resources that are often available. Quite often compute resource speed is a limiting factor of debug. The verification engineer creates their experiment, and submits it to the debug job dispatch queue. Half an hour later, the engineer returns to a completed simulation and continues the debug process. The intervening half hour may or may not have been utilized effectively, but quite often, it is not utilized on the same issue. The technique described here can be utilized to allow the engineer to parallelize their experiments. As soon as one job is kicked off, the engineer can begin creating the test environment for the next experiment and submit it along an independent revision branch on a parallel computing resource.

The resulting spreadsheet should be captured as a historical metric of the debug process that will travel along with the block or blocks of IP that were debugged. Other metrics that are useful to capture during this process are:

- Revision branch tag used to independently test each hypothesis or chain of hypotheses
- Pointer to the log files for each hypothesis
- Start time of issue debug
- Finish time of issue debug

Querying Coverage for Debug

Many verification metrics have unexpected, but valuable uses. For example, debug can often be facilitated by an innovative use of coverage metrics. One useful debug technique is to first find a testcase where the device is configured in the same manner as in the failing testcase, but is exhibiting the *correct* behavior. The engineer can find this testcase using functional coverage. First the engineer determines how the device in the failing testcase was configured. For example, the DMA block was configured to run with a granularity of 1 KB transfers. They then correlate functional coverage on the DMA granularity setting from the entire regression's aggregated functional coverage back to the testcases that created it. This creates a list of testcases that configured the device in the same manner. If a testcase can be located that passed while the device was configured in this manner, it can be used as a baseline to compare the behavior of the failing testcase.

The utility doesn't stop at functional coverage either. Any type of coverage metric can be used in the same manner. For example, consider Table 10.4.

Table 10.4 Coverage Usage for Debug

Coverage type	Usage example
Assertions	When was a given corner case exercised successfully?
Code coverage	When was a given module heavily exercised?
Software coverage	When was the device exercised in a given usage model?

This coverage correlation technique can be most effectively used with an efficient rerun scheme.

Debug Metrics

Like all other processes, debug itself creates metrics that should be tracked. Metrics tracked for debug tend to fall into the historical category of metrics that you hope you'll never need, but that you'll be really glad you have when you do need them. Among the metrics that should be stored are:

- Time taken to debug each issue.
- Number of reported issues per device feature.
- Number of issues reported per verification intellectual property package used.

Time Taken to Debug Each Issue
This metric can be used to check the efficiency of various debug techniques. It can also serve to point out blocks or block integrations that may be too complex and require refactoring.

Number of Reported Issues per Device Feature
This metrics can also be used to find portions of the design that might need extra attention. It can also indicate that there are portions

of the verification environment that need extra attention depending on how the issue was ultimately resolved.

Number of Issues Reported per Verification Intellectual Property Package Used

This metric can be used as an indicator of the value of various VIP packages.

Summary

In this chapter, we have shown how metrics can be used to streamline many verification debug processes. We have shown only a few examples. However, even these examples can reap great gains in productivity. There are many other opportunities for streamlining debug processes using metrics. The user is urged to explore!

Part III
Executing the Verification Process

Chapter 11
Coverage Metrics

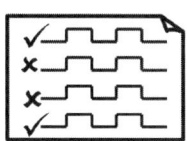 

Introduction

There are several different coverage metrics that will be used to track each of the verification technologies described in this part of the book to closure. Because several of the metrics are used for more than one of the technologies, their descriptions have all been summarized here at the beginning of Part III.

Functional Coverage

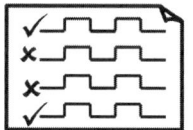
Functional coverage is used to measure the number of interesting scenarios that the device under test has been simulated in. It is a modeled form of coverage and as such, it must be implemented by a verification engineer. Functional coverage design is derived directly from the features, events, and attributes laid out in the verification plan.

Let's look at functional coverage in the context of the collaborative planning process described earlier in the book. The planning process consisted of a few basic steps. They were:

- Describe a feature of the device.
- Describe how to detect that the feature has been exercised.
- Describe attributes of the device are interesting to measure in conjunction with the feature being exercised.

For example, we may be interested in the snoop feature of a processor. First, we describe the feature in generic terms.

Snooping
```
The processor has a snoop signal that allows
other devices in the system to request memory
whose latest copy is contained within the cache
of the processor. The processor can respond to
a snoop signal by doing nothing, or by
flushing the contained memory back out to the
system memory.
```

Next, we describe how to detect a snoop has been issued to the processor.

```
A rising edge on the 'snoop' input signal
starts the snoop state machines within the
processor.
```

Because the feature will cause instructions to fetch their operands from different locations, we'd like to know that each instruction has been executed in the presence of an external snoop signal. The timing of these events is shown in Figure 11.1.

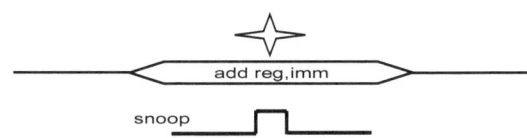

Figure 11.1 Snoop Signal Coincident With Instruction Executes

We might also like to know that each instruction has been executed while accessing each address of contained in our 1-bit address space (yes this is a toy example!). So, our final description of what we would like to measure is:

```
On the execution of each external snoop, we'll
measure  the  instruction  executed  and  the
address that each of its operands accessed.
```

All that remains is to code the functional coverage group from our verification plan. The example below is coded in the *e* language. A similar implementation could be performed in SystemVerilog.

```
event snoop is rise('top.snoop')@clk;
cover snoop {
    item instruction.opcode
    item instruction.addrs1
    item instruction.addrs2
    item instruction.addrt1
};
```

The preceding cover group description will measure and record the instruction op-code, its two source addresses, and its target address when each snoop signal is detected in the execution engine. The output results of the cover group would look something like those shown in Table 11.1. The count indicates the number of times that the triggering event has been observed for the attribute values shown in that row.

Table 11.1 Functional Coverage Results

Op-Code	Src Addr1	Src Addr2	Tgt Addr1	Count
Add	0	0	0	23
	0	0	1	1
	0	1	0	2
	0	1	1	2
	1	1	1	0
Sub	0	0	0	0
	0	0	1	1
	0	1	0	25
And So On				

By viewing our functional coverage results, we can tell whether or not the snooping feature has been exercised in every manner that we're interested in.

Even better! Using a constrained random testing methodology with functional coverage as a closure metric as shown below, we can get out of writing some of the testcases required to exercise this feature.

Code Coverage
Code coverage simply measures the lines of RTL code that were executed by the simulation. This is implicit or automated coverage and is often called "implementation coverage." Most modern simulators include a code coverage tool.

By using both functional and implementation coverage a set of checks and balances can be set up. Table 11.2 shows the matrix that can be used to draw inferences from these two sets of data.

Table 11.2 Implementation vs. Functional Coverage

	Low-functional coverage	High-functional coverage
Low-implementation coverage		
High-implementation coverage		

A combination of low-implementation and low-functional coverage may indicate that the project is in its bring-up phase. There simply isn't enough simulation infrastructure in place to provide adequate coverage in either of these spaces.

High-functional coverage results in conjunction with low-implementation coverage results could indicate:

- An incomplete functional verification plan
- Blocks of the design that are not used

High-implementation coverage in conjunction with low-functional coverage could indicate:

- There are blocks of the design that are unimplemented

High-implementation coverage in conjunction with high-functional coverage is fundamentally a good indicator. However care should be taken to review the verification plans and perform more random simulations. There are of course opportunities for false positives here. If holes in the verification plan happen to coincide with unimplemented blocks good implementation coverage and high-functional coverage may still be achieved.

Test-Based Coverage

Test-based coverage is related to directed testing. In a directed testing strategy, the verification plan consists of a series of directed testcases that are to be written to exercise the device in the manner that verification stakeholders are interested in. Coverage is considered complete when all the testcases have been written and are passing. Some tools are currently available to automatically track the implementation of directed testcases against a verification plan in this manner.

Assertion and Checker Coverage

Most assertions consist of two clauses, an initial clause that activates the assertion and a test clause that specifies what is to be evaluated once the assertion is activated. While failures should trend toward zero, every dynamic assertion should be activated. In this fashion, dynamic assertion activations serve in the same sense as functional coverage of input stimulus. If an assertion has not been activated, then the stimulus required for the activation has not been introduced to the device.

Chapter 12
Modeling and Architectural Verification

Introduction

In larger, more complex projects, such as the massively integrated SoC designs commonly produced today, exploration of various architectural solutions is often performed first. The purpose of architectural explorations is to analyze the tradeoffs between different possible architectures of the chip based on criteria such as performance and power consumption.

These explorations are typically done using transaction-level modeling (TLM). By modeling the device at the transaction level, enough of the device complexity is removed so that simulations with different architectural parameters can be performed easily. Transaction-level modeling is often done using SystemC.

TLM can be executed at a variety of abstraction levels. The various abstraction levels provide trade-offs between performance and accuracy. The algorithmic level allows the user to test algorithms to be implemented by the device without regard to the performance of the device or timing information. Resources available to the device are also not taken into account here. This level of abstraction is simply used to determine if the algorithm to be performed is feasible and produces correct results.

The programmer's view adds the conceptual limitations of the memory spaces available to the device and any master or slave devices used by the algorithms. In this manner, the model takes into account limitations of the resources required by the algorithm. The architect can now make decisions about reimplementing the algorithm, or adding more or faster resources to the planned device.

The programmer's view plus timing adds information about the bus architecture. It adds general information about the amount of time required to access resources from the bus and models bus contention as well. This is where the first performance testing takes place.

The cycle-accurate level breaks transactions down into the atomic bus-level transactions that they are constructed from. The user can do accurate architectural studies of bus traffic. At this level, the power required to model these small transactions begins to erode the performance of the model in general.

Finally, the register transfer or RT-level models the device down to the signal and bit accurate level. The performance at this level is similar to the performance offered by RTL simulation. This model provides the most accurate results with respect to how the planned device will actually perform.

How to Plan

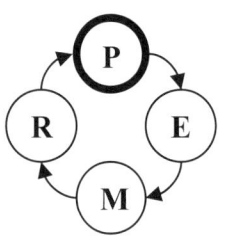

There are several aspects of architectural level verification that might be important to verify. Some of these are:

- Algorithm correctness
- Necessary performance
- Memory footprint
- Bus congestion

Planning and the Project Stakeholders

The key stakeholders for planning architectural verification are:

- System architects
- Verification engineers

Other stakeholders that can benefit from attending the planning session are:

- Design engineers
- Firmware engineers
- Application engineers

These stakeholders benefit by being exposed at a very early stage to the design intent in its near original form.

Architectural verification may be performed by either the system architect or the verification engineer. The planning session should follow the collaborative process outlined in the chapter on verification planning. Architects should describe each feature of the device, how it should be exercised, and what aspects of the feature are important to verify. A performance or behavioral check should be identified for each feature.

One of the key benefits that the verification engineer can offer is insight into how to make the architectural verification constructs described in the planning session portable to the other abstraction levels of the project.

Tracking to Closure

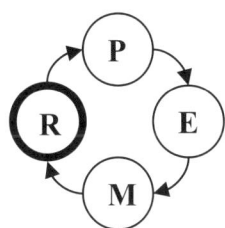

It is important to track that the architectural models have been suitably exercised using functional coverage and that they have been suitably verified using assertion and checker coverage. It is of even more importance to verify that these architectural requirements are still met as the design is implemented in hardware. Using a suitable reuse methodology we can easily deploy the coverage and checking metrics described here in our hardware verification environments as well.

Reusing Architectural Verification Environments

With some up-front planning, we can reuse the following aspects of our architectural verification environments throughout the project:

- Transaction stream generators (also known as bus functional models (BFMs) or scenario drivers)

- Correctness checkers and assertions
- Functional coverage
- Checker and assertion coverage

The architectural pattern used (verification environment architecture in this context, not device architecture) is the proxy pattern familiar to object-oriented programmers. The basic principle of the proxy pattern is to build our complex operations on abstract inputs so that they can be easily reused. Figure 12.1 illustrates this concept.

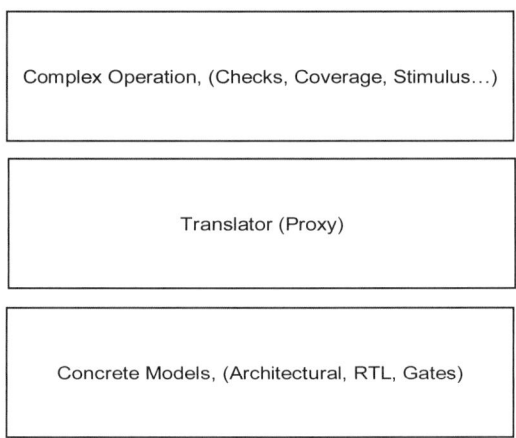

Figure 12.1 The Proxy Pattern

For more information and specific examples on how to implement this pattern there are several good reuse methodology volumes available including the Incisive Plan to Closure Methodology Manual from Cadence Design Systems.

Summary

In this chapter we briefly explored the application of metric-driven techniques to architectural verification. We described the various stakeholders and how they should be involved in planning. We also pointed out the reuse concerns for ensuring that architectural concerns continue to be met as the device moves from an architectural concept to production ready hardware.

Chapter 13
Assertion-Based Verification

Introduction

Assertion-based verification is the first chance that the design and verification teams have to verify the functionality of the design vs. the functional specification and the implementation specification. While assertion-based verification can be very simple, it is one of the best opportunities to make sure that the verification project is effective and completed on schedule.

Properly executed assertion-based verification environments provide the following benefits:

- Assertions provide a formal means for designers to encapsulate their knowledge about their design intent at a higher abstraction level than the implemented RTL.
- Assertions reduce debug time throughout the project.
- Assertions ensure smooth integration of design blocks.
- Assertions can be used to prove that portions of the design work as specified in the assertion.
- Assertions catch bugs early in the project that will become exponentially more expensive to debug as the project proceeds.
- Assertions can be reused from the block all the way to the system level. Starting in static, formal environments, moving into dynamic simulation environments, and finally into emulation/acceleration environments.
- Assertions provide coverage of important internal design states.

Assertions are Boolean and temporal checks that monitor signal values from the DUV. Assertions can be written in a number of languages including Verilog, SystemVerilog, *e*, VHDL, and PSL.

The *e* language assertion in Figure 13.1 demonstrates a simple Boolean check. This assertion checks that a grant is always received within four cycles of a request. The Boolean expression is activated and evaluated each time the event "clk" is detected as denoted by the "@clk" clause.

```
expect req_after_gnt is @req => {[0..4]; @gnt}
                @clk;
```

Figure 13.1 *e* Language Boolean Assertion

Assertions can be used for verification in one of two ways. An engineer may simulate their design with the assertions (dynamic assertion-based verification), or they may prove the correctness of the design using formal techniques (static assertion-based verification). In the case of dynamic assertion-based verification (ABV), the assertions will check for proper operation during the simulation.

Using formal analysis techniques, assertions can actually be proven to always be correct. This is a very powerful technique because it guarantees that the functionality checked by the assertion is correct. These formal proofs are best leveraged on relatively small portions of the design that contain state machines and control logic.

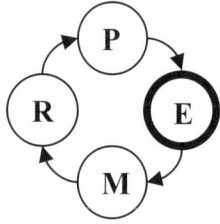

While formal verification was once the domain of the Ph.D. formal analysis expert, recent advances in technology have made this technique accessible to everyone. Engineers simply submit their synthesizable design and assertions to the formal verification tool. These tools, such as IFV from Cadence Design Systems, will attempt to prove that the assertion can never be violated. They will generate one of three answers. If the assertion can be proven, the tool will simply declare that the assertion was proven. If the assertion can be

13 Assertion-Based Verfication

violated by the DUV, the tool will issue an error message and display a set of waveforms that illustrate the sequence of signals that induce the design to violate the assertion. Finally, if the tool cannot prove the assertion in an allotted amount of time, it will issue a warning stating that it could neither prove the assertion, nor find a counter-example that illustrated the design violating the assertion.

Formal ABV requires the use of a second class of assertions called properties. Formal proof engines work best on relatively small well-constrained blocks of logic, such as those found at the block level of a design. To keep the problem that the formal engine has to solve well bounded, it is necessary for the engineer to define boundary conditions at the edge of the module. These boundary conditions are called properties and are declared using the same assertion constructs that are used for the assertions that are to be proven.

Formal ABV is of particular value because it allows the design engineer to begin verification before any testbench apparatus is available (Figure 13.2).

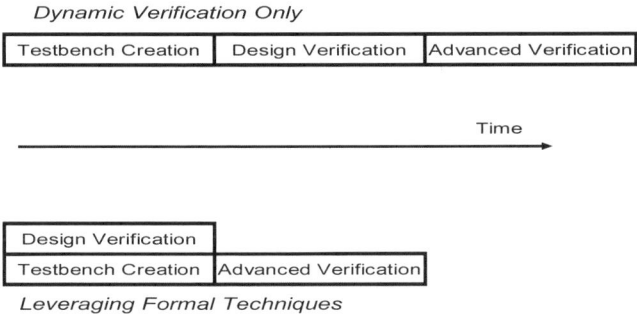

Figure 13.2 Formal Schedule Savings

Without formal techniques, the design engineer must either create a testbench, or wait for the testbench that the verification team will create for advanced verification, before any testing of the design can be done. Using formal techniques, the designer can begin checking the design as soon as a synthesizable module is available.

The sooner a bug is found the easier it is to debug. Several studies have found that the amount of time required to debug an issue increases exponentially with the amount of project time that passes before the issue is detected. By performing block-level verification before a testbench is available using formal techniques, many issues can be corrected immediately by the designer before they are detected by more costly simulations.

How to Plan

Extreme Programming and Design Assertion Planning

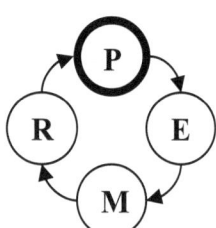

One concern that is frequently raised when planning design assertions is: "How do we know that we wrote assertions that correspond to *all* of the behavior of the device?" In other words, how do we know we have a complete set of checks for the features implemented by the design block?

The answer is actually quite simple. Taking a page from software engineering, we'll use an extreme programming concept. Put concisely, this is an edict that says:

"No implementation will be performed until a testcase exists for the planned implementation"

As implementers determine what features to implement and how to implement them, they first determine how those features will be tested. Only after implementing the testcase, they begin implementation of the feature itself. In the case of hardware-based design, we'll use assertions instead of software testcases to verify each piece of hardware functionality. When all our assertions pass, we'll have our first level indication that the design is ready to go. No more subjective estimations: if the assertions pass, then every feature has been implemented and tested.

There is a second powerful edict of extreme programming. It is:

"Nobody programs or tests alone! Always take a buddy!"

It's simply too tempting to dive straight into implementation with no testing. The implementation seems short. What could go wrong? Just bang out the code and get it done, right? Never mind that "verification takes 70% of project effort" quote you keep hearing in the hallways. Everything will be fine!

With a development partner, the temptation to skip the rules is highly diminished. Two engineers are much more likely to follow a system than one. It's just human nature. You'll have more fun.

There is a second advantage to having a codeveloper. It might seem really sexy right now to be the "sole" developer of that shiny new hardware block. It might seem to imply importance and job security if you're the only one that knows the blocks inner-workings. But, especially with today's bent toward reuse, when you get the 500th support call from yet another engineer that just picked up your hardware block, the gleam will definitely be off the rose.

For assertion-based verification planning, we're going to work two at a time and we're going to identify an assertion for each feature before we implement it. Let's see what this process looks like in practice.

We'll start with some raw material, an empty HDL module, some form of design intent, and an empty executable verification plan. As we declare functionality, we'll begin to declare the registers, inputs, outputs, and wires that are needed to implement it. We won't go any further on implementation though. Before we implement, we'll write the assertion that verifies the functionality we're about to implement. We define the necessary signals so that at any given moment, the design shell and its associated assertions should compile and execute in a simulator or formal proof engine.

As we're writing our assertions, we will add descriptions of them to a hierarchical executable verification plan. In this manner, we automatically create the tracking mechanism that we will use to tell ourselves and the rest of the team that the design is ready. In addition, we're creating a document that will tell future users of the device exactly what is tested. This documentation will come in handy for debug at higher levels of integration as well. As we kick off simulations, we'll track results back into the executable verification plan. When the verification plan says we have 100% coverage of assertions, our design is complete.

Let's look at an arbiter for example. First, we construct our empty arbiter module as shown in Figure 13.3.

```
module my_arbiter{
begin module;

input reset;
input clk;
input req1;
input req2;
input req3;
output grant1;
output grant2;
output grant3;

end module;
}
```

Figure 13.3 Arbiter Module

Our first feature for the arbiter will be its reset behavior. We know that when the reset signal transitions to low that all grant signals should be forced to their de-asserted state. We can write an assertion to check this:

```
my_assertion: assert (GNT1 && GNT2&& GNT3 && !RST);
```

And finally, we'll add the feature to the verification plan (Figure 13.4).

Now when we run our simulation, all of our assertions will of course fail and we'll get an executable view of our verification plan like Figure 13.5.

Arbiter

Reset

The grant signals should all be de-asserted when reset is asserted.

```
/sys/arbiter/reset_assert
```

Figure 13.4 Verification Plan Source

Arbiter

Reset – 0%

The grant signals should all be de-asserted when reset is asserted.

```
/sys/arbiter/reset_assert - 0%
```

Figure 13.5 Verification Plan Executable View

The flow just described is shown in Figure 13.6.

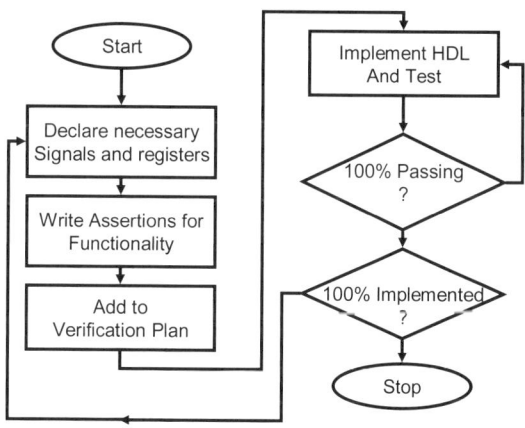

Figure 13.6 Assertion-Based Planning Flow

As we implement the functionality, we'll see the 0% indicator move to 100%. By repeating this process for each piece of functionality, we are guaranteed to have assertions corresponding to each operation of the device. In fact, the assertions actually define the operation of the device.

The verification plan and assertion library will grow after each implementation round. As the plan begins to grow we might see Figure 13.7.

Arbiter

Reset – 50%

The grant signals should all be de-asserted when reset is asserted.

```
/sys/arbiter/reset_assert - 100%
```

The grant signal must assert within 15 cycles of the assertion of the request signal.

```
/sys/arbiter/gnt_req - 0%
```

Figure 13.7 Executable Verification Plan

The plan will continue to grow in this manner until all behaviors of the module have been tested and implemented. Notice the word order there, it's important. Tested and then implemented, not implemented and tested.

Where Do We Go From Here
Using the planned assertions described here, the design team can be confident that their creation plays well. Now it's time to make sure it plays well with others by handing the design over to the functional verification and integration teams. These teams begin their tasks with planning. Planning at this level is a collaborative effort between the design, verification, and software engineers for the device. The design engineer presents a block diagram of their design and

describes the features of the design. Each stakeholder (designers of adjacent blocks, integration engineers, verification specialists, firmware developers, etc.) then specifies how they intend to use the feature and what they need to see verified to feel confident the feature is implemented correctly.

The basis for the conversation is the feature set identified by the designer. Using the planning methodology shown here, the design engineer has a well-documented set of features which he can use to lead the discussion. Not only do they have an exhaustive list of features, but proof that each of the features has been exercised and verified. This can make verification planning sessions much more productive and efficient.

Tracking to Closure

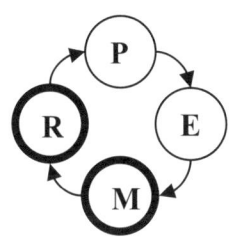

The tracking process for design-based assertions follows closely from the planning process. The basic metrics that should be tracked are:

- Number of assertion failures
- Number of assertion activations
- Number of complete proofs
- Total count of assertions

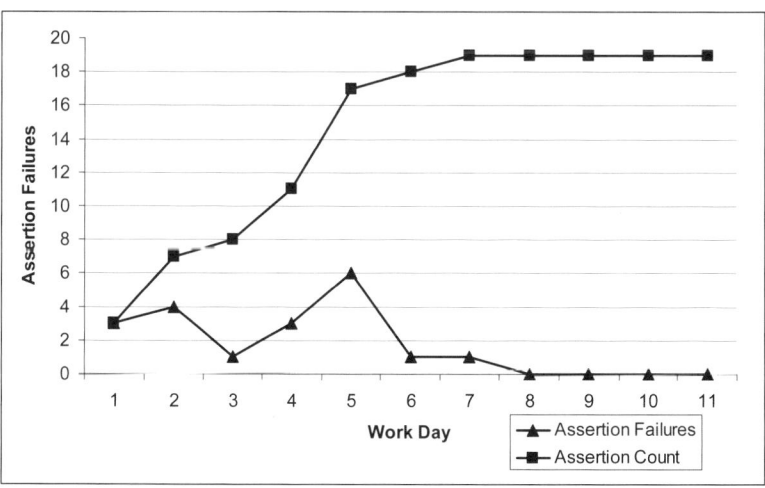

Figure 13.8 Assertion Failures and Count

If the planning methodology outlined above is followed, then the initial number of assertion failures should be high at the beginning of each implementation iteration and then trend downward as hardware is actually implemented. An example of this is shown in Figure 13.8.

Metrics can be used to assure that the proper "test *then* implement" methodology is being followed. We generate a report like the one in Table 13.1 that shows the number of failures detected per assertion.

Table 13.1 Errors Detected By Assertions

Errors detected	
Assertion name	Error triggers
ar_bandwidth_ch	2
Dn_FIFO_Empty	0
Dn_FIFO_RW	3

If the suggested methodology is followed, all assertions should initially fail because no functionality is implemented. It can be seen that the dn_FIFO_Empty assertion either checks nothing, or was added after the initial code was implemented. Keep in mind that later in the project this may indicate that unexpected bugs were found and new assertions were added to provide more detailed checks for these issues.

If the assertion failure number does not initially start high, it is an indication that the planning methodology is not being followed.

Assertions are only valuable in dynamic simulation if the design is properly stimulated. For this reason, when assertions are used in this "dynamic" mode, it is important to link coverage of their activating event into the executable verification plan. If less than 100% assertion activation is achieved, the project team should continue to generate stimulus to activate the uncovered assertions.

The formal proof results should be tracked as well. The team should expect to see all formal proofs trend to the proven state. In the event that counter-examples are detected, the engineering team should fix either the DUV or the assertion if it contains errors. In the event that the third inconclusive state is encountered, the engineer has two choices. First, they can instruct the formal proof engine to expend more effort proving the assertion. If the proof results are still inconclusive, then the assertion should be utilized as a dynamic assertion in conjunction with functional coverage to make sure that the assertion was sufficiently exercised.

Most assertions consist of two clauses, an initial clause that activates the assertion and a test clause that specifies what is to be evaluated once the assertion is activated. While failures should trend toward zero, every dynamic assertion should be activated. In this fashion, dynamic assertion activations serve in the same sense as functional coverage of input stimulus. If an assertion has not been activated, then the stimulus required for the activation has not been introduced to the device.

Some formal assertion tools provide built-in assertions that are common to all hardware designs and point out such design issues as intrinsic race conditions. Another simple metric is to ensure that these tools have been used on each design block.

Formal tools also offer facilities that report how many of the design inputs, outputs, and logic are not touched by defined assertions. This gives a measure of the completeness of the assertions defined for the design.

Opportunities for Reuse

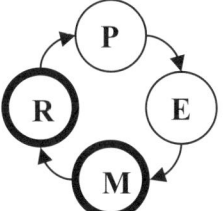

Assertions can be used at all levels of verification from architectural studies to block-level verification to system-level integration. If the assertions are properly constructed, they can be significantly reused from level to level.

Much has been written about assertion reuse. There are several references in the bibliography on this subject. For our purposes however, it is important to track how effectively assertions in the project have been reused. Using this information, the project team can adjust their methodology to make more effective reuse of available verification IP. The verification project manager should track how assertions were reused between the architectural, block, integration, and system level of verification. This can be done by using a standardized assertion naming scheme and then tracking which assertions found errors at each level of verification. This metric illustrates not only which assertions were reused effectively, but also which assertions were most effective in pointing out design issues. An example of an assertion tracking metric table is shown in Table 13.2.

Table 13.2 Assertion Metrics

Assertion Catalog		
Assertion Name	Verification Level	Errors
ar_bandwidth_ch	Architectural	2
ar_bandwidth_ch	Dynamic Simulation	1
ar_bandwidth_ch	Dynamic Emulation	1
dn_FIFO_Empty	Formal	0
dn_FIFO_Empty	Dynamic Simulation	2
dn_FIFO_RW	Dynamic Simulation	3
dn_FIFO_RW	Dynamic Emulation	1

This raw metric data can be consolidated and analyzed to yield results such as the reuse table (Table 13.3).

Table 13.3 Assertion Reuse Metrics

Assertion reuse		
Assertion name	Levels used	Errors
ar_bandwidth_ch	3	2
dn_FIFO_Empty	2	0
dn_FIFO_RW	2	3

Alternatively, reports might be created that show where the largest opportunities lie for improvement. Such a report is given in Table 13.4.

Table 13.4 Assertions Used Only Once

One use assertions	
Assertion name	Levels used
ar_intrpts	1
dn_opcode	1
dn_onehot	1

Summary

We have seen how assertions can be leveraged to jumpstart our verification efforts. A new planning methodology that leverages designers was outlined and it was shown how the output of this planning can be taken as the input for the feature-based verification planning described earlier in the book. We also saw how to track metrics from assertions to gauge project status and also to gauge reuse opportunities.

In Chapter 14 we'll look at metric-driven simulation-based techniques.

Chapter 14
Dynamic Simulation-Based Verification

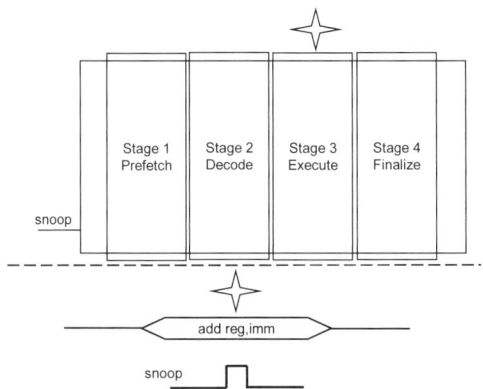

Introduction

This is the classic bread and butter of functional verification. Mention functional verification in most semiconductor companies and simulation-based verification is what comes to mind.

Conceptually, this is one of the simpler flavors of verification to grasp. An application called a simulator runs on a workstation and allows the user to simulate a device as it would actually work in a real system. The user can describe dynamic stimulus that is to be applied to the device and then monitor the response of the device as the stimulus is applied over time. The user can specify signals that are to be driven to produce the stimulus. The user also has access to the values driven on each signal within the device at any given time during the simulation. Using values of these signals, the user can check for proper operation of the device under verification.

A design engineer creates a model of the device in RTL (register transfer language) using a language such as Verilog or VHDL. Both languages offer procedural programming constructs similar to those available in C. They can therefore be used not only to create the model of the device, but also to create the testbench that will be used to test the device. Verilog and VHDL are both structured languages that allow models to be divided into subsystems. This ability to organize models into subsystems encourages reuse of both designs and their testbenches.

In addition to Verilog and VHDL, most simulators provide interfaces to verification-specific languages such as SystemC, e, and SystemVerilog. These languages offer various levels of object and aspect orientation that facilitate reuse of verification intellectual property (VIP).

Simulation-based verification systems typically mirror the implementation of the design. For example, if a design is built as a number of design blocks, and the blocks are then integrated to the chip level, there will typically be an independent verification environment for each design block, and an integrated verification environment that tests the integrated chip as a whole (Figure 14.1).

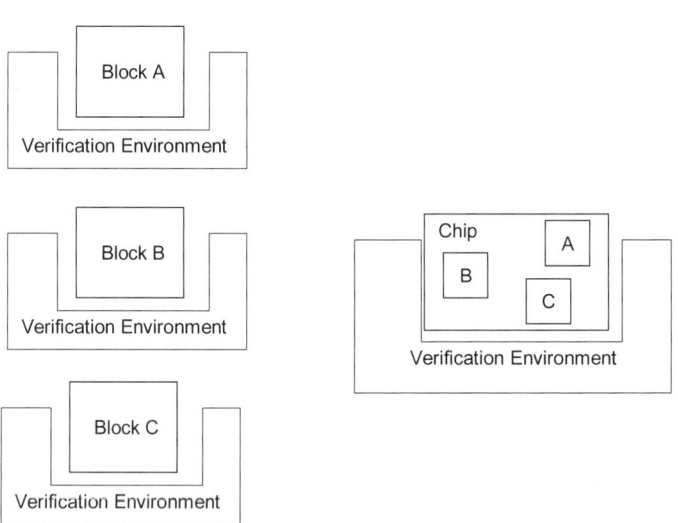

Figure 14.1 Block and Chip Verification

How to Plan

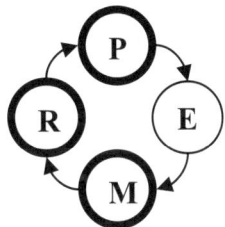

Planning for simulation-based verification uses the collaborative techniques described in Part 2. To reiterate here, a feature-based planning approach should be used to determine what coverage metrics should be collected from the verification environment. In simulation-based verification, three types of coverage are widely used. They are:

- Functional Coverage
- Code Coverage
- Test-based Coverage

Planning and the Project Stakeholders

The stakeholders that should be involved in planning for simulation-based verification are:

- Verification engineers
- Design engineers
- System architects
- Firmware engineers
- Application engineers
- Design lead and/or manager
- Verification lead and/or manager

Each of these stakeholders is important first for the perspective they can offer to a holistic understanding of the device under verification as described in the chapter on verification planning. Remember, design intent is translated differently by each stakeholder, and the planning session is our opportunity to align these translations (Figure 14.2).

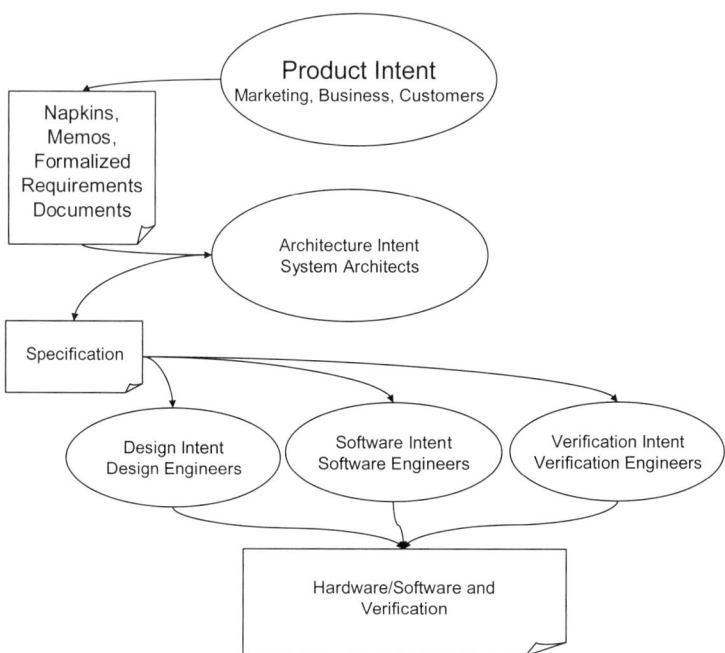

Figure 14.2 Translations of Device Intent

In addition to contributing their unique viewpoint to the team, each contributor needs to specify and discuss which aspects of each feature are important to them. In some cases, they will help to define metrics that will objectively measure verification closure vs. their concerns. In other cases, they will note metrics that have already been defined and add them to their individually defined "view" of the verification plan. Let's take a look at some of the concerns each stakeholder might have.

Verification Engineers
Verification engineers lead the planning process. They must be concerned with every aspect of verification and their "view" into the verification plan should encompass *everything*. The verification engineer should also elicit input from all the other contributors. A verification engineer's constant litany during the planning process should be:

- "What does this feature mean to you? How will you use it?"
- "In how many different ways can this feature be configured? How is it configured?"
- "How can we detect when that aspect of this feature has been exercised? How can I build an event to detect that?"
- "What is important to measure about this aspect of this feature? Should that value be captured when we detect the feature has been exercised or at some past or future event?"
- "How many of the values must be observed for you to feel comfortable that this feature has been completely verified?"

Design Engineers
The design engineer possesses knowledge of how the device has *actually* been implemented. As such, they present the description of each feature of the portion of the device that they were responsible for implementing.

While their hardware enables the more complete chip-level behavior that is of interest to most of the stakeholders, the design engineer's concerns may be more focused. They may want to ensure that all their code was actually exercised using code coverage as a metric. There may be specific complex scenarios that they know will exercise corner cases within their design. The verification of these scenarios might be gauged using functional coverage.

If the design engineer has used assertion-based verification techniques as described in this book, they should enter the verification planning session with an assertion-based plan of their own. The design engineer should watch for opportunities to apply preexisting assertion coverage as a useful metric for other stakeholders' concerns.

System Architects
The system architect may have produced the earliest translation of the original intent for the device when they defined the system-level specification. They should be particularly observant during the planning session to make sure that the intent translations expressed by all other stakeholders support the original design intent. They

should be able to clarify understanding of design intent when stakeholders have translations that conflict.

System architects should look for opportunities to apply any system-level modeling metrics to the rest of the verification process. Performing architectural verification to qualify the initial design partitioning decision is a good first step on the way to a high-quality product. However, it should also be verified that the real device provides the performance and capabilities that the architectural level models assumed. Often, assertions, functional coverage groups and behavioral checkers that were created for architectural verification can be reused directly in simulation-based verification.

Firmware and Application Engineers
Software engineers are responsible for creating the applications that will utilize the device. They can offer another translation of the design intent from the system view. They should carefully observe the planning session to ensure that the device is being implemented in a manner that will actually be usable.

The concerns of firmware and application engineers will vary based on the level of integration of the device. At the block level, they may only be concerned that each feature was exercised while configured in the manners that their software will eventually use. There may also be key scenarios of interest that they know their software will create in the block.

Concerns will vary with the context that each block will be used in as well. A software developer might not be interested in the internal communication protocols that are used by a DSP. However, the developer of the firmware for the USB port of a device might be very interested in checking that every possible USB transaction has been exercised and verified.

As the blocks are integrated into subsystems and eventually into the entire chip, more elaborate concerns will emerge. Scenarios may include the appropriate configuration of many if not all of the blocks in the chip, the specification of specific input transactions at the device's periphery, and the execution of firmware or application code.

Design and Verification Lead and/or Manager
The design and verification leads participate from a project management point of view and can also act as expert consultants and facilitators. They ensure that the planning session proceeds at an efficient pace and offer their leadership where necessary to focus the planning team.

Taxonomy of Simulation-Based Verification

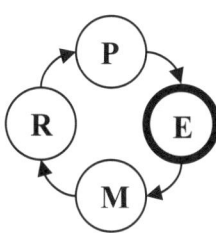

There are two basic types of simulation-based verification: directed testing and constrained random testing. These two types of testing are distinguished by how stimulus is generated for the device under test, how the output behavior of the device is checked, and how closure is tracked.

Directed Testing
In directed testing, the verification engineer creates deterministic descriptions of the testcase to be executed using one or more of several different verification languages.

The engineer is then responsible for writing a series of checks that determine if the device is operating correctly. There are generally two methods of checking.

Strengths
Directed testing for the most part is simple. It is the classic workhorse of simulation-based verification. It can be accomplished using standard in-line procedural programming techniques. Procedural, nonobject oriented, programming is a style of programming that is familiar to everyone.

While it may be time consuming to write testcases using this methodology, the task is almost always straight-forward.

Limitations

Directed testcases are by nature isolated from each other. The testcase tests only what it was intended to test (at best). There is little if any leverage with other portions of the device. Even if there is leverage, it is not intentional and often not perceived.

In one processor verification project that used directed testcases, a large set of testcases were created to verify the FPU. The FPU was functionally clean on first pass silicon and everyone declared success. About two weeks later, the testcase engineer received a call from the production testing floor. The test engineers wanted to know what the series of testcases labeled *fpu_xxx* did. As it turned out, this was the best set of test vectors the production floor had for detecting an issue with the instruction cache of the same device. The FPU testcases, simply because of an unplanned fluke of their architecture, were very well versed at thrashing the instruction cache of the processor. Had the verification team been aware of this they could have made use of the fact by running independent instruction cache checkers with the FPU tests. They could have at least gained respect from the production test team by flagging the *fpu_xxx* series as producing interesting instruction cache activity.

Another limitation of directed testing is that each aspect of the verification plan is targeted by a single testcase. The number of testcases that must be written grows very rapidly with increasing device functionality. This can lead to teams of testcase writers numbered in the tens or twenties. It also leads to regressions that contain up to tens of thousands of testcases. Directed testing is a labor intensive technique.

Constrained Random Testing

Constrained random testing can help eliminate much of the manual nature of directed testing. A verification apparatus is used that can randomly stimulate the device under verification and automatically check that the device behaved correctly based on the random stimulus applied. The application of random stimulus to your device might immediately bring the following three questions to mind:

- If the stimulus is random, how do I know what's been tested?
- How can my verification environment possibly check every random behavior?
- What keeps the randomly selected stimuli from being gibberish?

How Do I Know What Has Been Tested?

The first question leads us right back to metrics. The metric in this case is functional coverage. Functional coverage is used to detect specified scenarios and measure various device states when they are detected. By tracking the occurrence of these scenarios, we know what random stimuli are being applied to the device. Functional coverage will be covered in more detail in a following section.

How Can My Verification Environment Check Every Random Behavior?

Behavioral models are used to check the behavior of the device as it is stimulated. These models monitor the stimuli applied to the device under verification and then produce expected results based on the input stimuli. The expected results are compared to the actual results produced by the device. Figure 14.3 shows the architectural setup of these behavioral checkers.

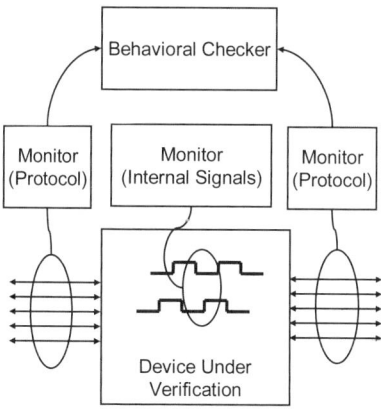

Figure 14.3 Behavioral Checking Architecture

There is one final metric-specific note on these checkers. As with assertions, it is important to be sure that the checker checked anything at all. In the assertion-based case, we simply made sure that all assertions activations had fired. We use the exact same technique to make sure that our simulation-based checkers are operating correctly. Each check is sensitized by some event that is detected by the verification environment. By observing the coverage for these events, we can ensure that the device under verification was exercised in a manner that sensitized each of our checkers.

Why Isn't the Randomly Selected Stimuli Gibberish?
The random stimulus can't be completely random. The space of illegal stimulus for a given device is usually much larger than the legal stimulus space. Constraints can be applied to limit the random generator to generate stimulus within the legal subset that the device can tolerate.

Care must be taken not to eliminate all illegal conditions however. For example, network routers are designed to detect both malformed packets and packets with bad parity. If the random environment is overly constrained then checks for correct device behavior in the face of erroneous external stimuli might never be exercised.

Strengths
Constrained random testing can eliminate most of the manual drudgery associated with using directed testing on too large of a project. It makes use of low-cost computing resources to automatically generate stimulus that would otherwise be generated by relatively expensive human resources. It also frees up engineers to begin the important work of debugging at an earlier time.

The behavioral checkers shown in Figure 14.3 are passive by nature. They only observe signals and events passed to them by the monitors (the monitors are passive as well). Due to their passive nature, these checkers can easily be included in verification environments that concentrate on other portions of the chip. For example, in the FPU example cited above, the instruction cache checkers could have easily been included to check for proper behavior.

Weaknesses
Constrained random testing can carry a steeper learning curve along with it. These systems are best implemented using object- and/or aspect-oriented programming techniques. Quite simply, there aren't as many engineering resources available that are well versed in these programming methodologies.

For a small design project, it might be prudent to pursue a directed testing methodology depending on the engineering resources available.

Tracking to Closure

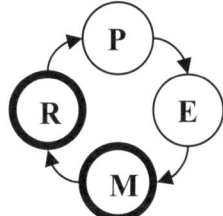

There are several different methods of tracking to closure using simulation-based verification. Each method corresponds to one of the methodologies for simulation-based verification described above. The simulation-based methodologies are:

- Directed testing with golden vectors
- Directed testing with self-checking testcases
- Constrained random testing

It is important to use all available metrics to track toward closure in a simulation-based verification project. Multiple metrics provide multiple perspectives into the project status and reduce the risk of missing key information.

Tracking to Closure Using Directed Testing

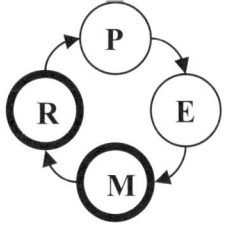

There are several metrics that are used to gauge the completeness of directed testing these metrics are:

- Testcase completion
- Code coverage
- Functional coverage

Testcase Completion

The first, most popular metric is simply to track the list of testcases that were specified in the test plan against the testcases declared completed by the testcase writers. This is a highly subjective and labor intensive tracking metric. Whether or not a testcase is complete is frequently a matter of human interpretation. The process is labor intensive, because someone has to gather the completion data from all the testcase writers.

Some automation and objectivity can be gained back by tracking whether each specified testcase passes or fails. Even this level of objectivity can fall victim to the possibility of false positives if it is used in isolation. Testcases that actually check nothing or exercise nothing will always pass.

Code Coverage

That brings us to the second metric which is code coverage. Code coverage is used as an independent metric that illustrates how much of the design was actually exercised by a given set of testcases. Using code coverage in conjunction with tracking testcase completion offers some balance to prevent false positives caused by misinterpretation of what each testcase is actually achieving.

Functional Coverage

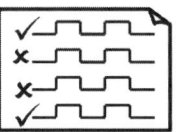
Functional coverage is sometimes used as a metric when working in directed testing environments. Occasionally it is used to gauge the advantages that can be gained by moving to a constrained random testing environment. Another use of functional coverage is to act as a failsafe mechanism for ensuring the continued efficacy of testcases in the face of design changes.

An example of this can be gleaned from the experience of a large x86 processor manufacturer in the 1990s. A team of thirty engineers had been assigned to custom design testcases to test the execution of various instructions in the presence of snoop signal on the external bus interface. The testcases were carefully crafted and timed to ensure that the instruction in question was executing exactly when the snoop signal arrived on the bus interface. This took quite a bit of work. The snoop signal had to be timed to arrive at the execution pipeline stage just as the instruction was executing there. The architecture is shown in Figure 14.4. The team of engineers spent a month crafting these testcases by hand and finished with a great sense of pride and sigh of relief.

A week later, the design of the processor was changed in such a way that all the timings were thrown off. The testcases were *testing nothing!* With functional coverage, the team was alerted to the problem immediately. They rewrote all the testcases and got the regression suite back up and running. Without functional coverage, they might have been using valuable regression resources to run testcases that did nothing.

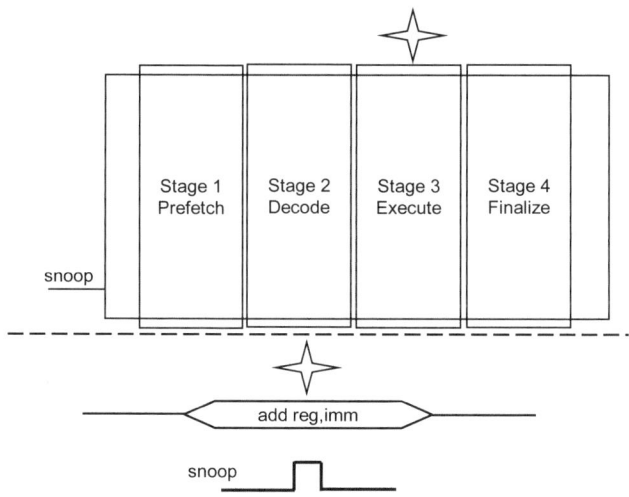

Figure 14.4 Timing Snoops

Tracking to Closure Using Constrained Random Testing

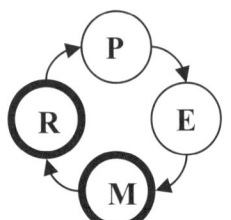

When using constrained random testing, there are a few key metrics to track. They are:

- Functional coverage
- Checker coverage
- Code Coverage

Functional Coverage

When tracking closure with functional coverage, the intended integration level of the functional coverage group should be taken into account as well. For example, let's assume we're verifying a device that will be segmented into the following integration levels:

- Block
- Unit Integration
- Chip Integration
- System Integration

As we specify each functional coverage group, we should also specify a coverage goal for that group in each of the integration levels. For example, the coverage of every available bus transaction for a given block might have the coverage goals given in Table 14.1.

Table 14.1 Coverage Goals vs. Integration Level

Coverage goals for block X bus transactions	
Integration level	Coverage goal
Block	100%
Initial integration	75%
Chip integration	30%
System integration	10%

At the block level, we have the most controllability and observability. We want to make sure that all the functionality is verified here where debug is the easiest. Coverage for all available bus transactions should be 100%. In addition to checking the block for use in our present device, we are also ensuring that the block supports all available bus transactions in case it is used in another project.

At the integration level, our goals change. It was important to verify all available bus traffic for the block. However, the block that it is integrated to Block X does not support one of the transaction types. Our coverage goal has been adjusted accordingly. We only expect to see 75% coverage at the initial integration level.

At the chip integration level, we are now confident that the interface to the block is fully verified and we merely want to see that the block participated to some degree in chip-level simulations. Here, the transaction coverage gives us a warm, fuzzy feeling that everything is OK, but it is no longer essential.

Finally, at the system level we are even less concerned with the exact transactions that are sent to the block in question. However, we want to make sure some traffic is being sent to the block to ensure that our bandwidth performance measurements are realistic.

Checker Coverage
It's not enough to track just functional coverage. 100% functional coverage does not guarantee that every feature was actually properly checked. In fact, each feature can be exercised and never checked. Once again, it's important to track multiple metrics. In addition to specifying functional coverage that indicates that each feature has been exercised sufficiently, we also need to specify and implement coverage that ensures each feature has been checked by one or more checks in the verification environment. As functional coverage ramps up to meet our goals, checker coverage must be tracked for completeness as well. A divergence in these numbers indicates that progress is not being made in either the area of stimulus or implemented checks.

Finally, it is important to check all issues detected against the existing functional coverage implementation. The constant question must be: "would we have found this bug had we reached 100% coverage?" Often times in constrained random environments, the answer is no. That's a good thing! The reason for using constrained random stimulus was because it can explore state spaces that you may not have thought of ahead of time. When an issue is exposed by "unplanned" stimulus the verification plan should be updated. The team should plan and implement an appropriate functional coverage group that would have indicated an incomplete verification effort had the random stimulus that exacerbated the bug not been created.

Code Coverage

The same considerations that were outlined for code coverage vs. functional coverage in the directed testing section above are important here. Code coverage and functional coverage independently are necessary but not sufficient metrics. When paired as described above, they create a very effective system of checks and balances for each other. This system provides more value from our constrained random verification environment.

Summary

In this chapter we explored using metric-driven techniques to augment the simulation-based verification process. We looked at the differences between directed test verification and verification using constrained random stimulus. We described the three key metrics used to track closure of simulation-based verification:

- Code coverage
- Functional coverage
- Checker coverage

We also looked at how each of the stakeholders should contribute to the verification planning process.

In Chapter 15, we'll look at system-level verification using acceleration and emulation technology.

Chapter 15
System Verification

An exciting new area in metric-driven verification is its application to system-level verification. The following chapter by Jason Andrews outlines how available technology can be used to apply proven metric-driven techniques such as coverage-driven verification at the system level.

Jason Andrews is a project leader at Cadence Design Systems, where he is responsible for hardware/software coverification and methodology for SoC verification. He is the author of the book "Co-Verification of Hardware and Software for ARM SoC Design" and holds a bachelor's degree in electrical engineering from The Citadel, Charleston, SC, and a master's degree in electrical engineering from the University of Minnesota, Minneapolis.

Coverage-Driven Methodology for Verification of SoC Hardware and Software Corner Cases

Introduction

One of the most difficult challenges in SoC verification today is determining how to make sure the hardware and software work together at the SoC level. Hardware verification has advanced to the point where the verification of individual functional blocks in a design can be achieved with reasonable confidence using constrained random testbenches, code coverage, assertion coverage, and functional coverage. Challenges remain in making sure the blocks work correctly when placed in the context of the SoC. On the other hand, the concept of embedded software verification is mostly nonexistent in SoC projects today. The primary way to find out if the software works with the hardware is to just run it and watch what happens.

The result is a commonly deployed three-step process for SoC verification:

- Perform comprehensive verification of each functional block of the design using:
  o A verification plan
  o Advanced simulation techniques
  o A farm of workstations for parallel simulation
- Assemble the blocks together to form the SoC:
  o Run some basic tests to make sure peripherals can be accessed, connectivity is good, and there is no contention on busses or interfaces
- Execute Software on the SoC:
  o As much as possible before committing the design to fabrication
  o Using emulation or FPGA prototyping to provide the needed performance

Unfortunately, this process results in many products that have subtle problems, most of which are caused by corner cases between the hardware and software. Many of today's consumer electronics are examples of this. They contain the most advanced features constructed with complex hardware and software, but suffer from periodic lock-ups or require periodic reboots. From wireless routers to Bluetooth mobile phones the story is the same, there are some conditions where the hardware and software hit functional corner cases and the failures occur. In the end nobody really knows what is happening because consumers have no way to debug the problem. Technical support may offer new firmware that may or may not address the exact problem, and most of the design engineers have gone on to the next project and are not interested in such subtle problems because they are very hard to find and fix and may not even be functional problems, but could be mechanical, electrical, or manufacturing issues.

This paper proposes new methodology to improve the three-step SoC verification process. The aim of this methodology is to produce higher quality designs by exposing the hidden corner cases that are not being found. As anticipated, the key is to apply additional stress to the boundary conditions of the design, but do it by including the embedded software in the process. An ARM926 PrimeXsys Platform SoC is used to demonstrate the proposed methodology and results are presented.

Coverification Defined

During the last 10 years the term coverification has been widely applied to any verification technique that included both hardware and software in an attempt to make sure each works with the other before designs are committed for fabrication. Today, nearly all SoC projects understand the benefits and readily admit that this is an area of struggle for which there is no easy solution. In fact, there have been many new products introduced in an attempt to address the problem, but each with a set of pros and cons that has limited any one technique from emerging as dominant. Engineers have been left with a daunting challenge of deciding which of the products and techniques to apply and when in the project is the best time to do it. Using multiple techniques in a divide and conquer approach has produced some progress, but not enough to declare victory when it comes to SoC verification. Attendees interested in a complete landscape of all of the products, techniques, and history should refer to [1].

Coverification is a term that is not the best fit for the activities that have been done so far. When engineers try to make sure hardware and software work together it is likely they are performing cosimulation. Historically, cosimulation meant the connection of two simulators where one simulator executed the hardware design and another executed the software. As many different techniques were developed for execution it's more general to think of *Cosimulation* as the execution of software together with the hardware, even if there is only one execution engine involved. When engineers try to figure out what is wrong when a failure occurs they are doing codebugging. *Codebugging* is the process of starting and stopping the system and observing the state of both the hardware and the software to understand where the problem lies. Although these activities are important they are not true coverification, but serve as a foundation on which coverification takes place.

Verification is the process of determining a design meets requirements. In practice, verification is the process of identifying and removing as many functional bugs in the hardware and software as possible. The oldest form of verification is to build it, run it, and watch what happens. Today, manual techniques such as visual inspection have been replaced by automated verification plans containing a set of goal metrics that are used to measure progress toward verification completion. By definition of the plan, if these metrics are achieved the design is verified. In hardware verification the process of verification planning to define the corner cases that need to be hit and the use of automated, constrained random stimulus to hit these corner cases is known as coverage-driven verification (CDV). To perform CDV the corner cases are converted into coverage points and the goal is to reach 100% coverage. The combination of using random generation to reach coverage points also results in new corner cases that engineers did not think of. Considering the wide adoption of CDV for hardware verification it is logical that Coverification should have a new definition that is specific to the verification problem. *Coverification* is the use of automated, constrained random stimulus and functional coverage metrics applied to the hardware design, the embedded software, and the combination of hardware and software. Performing coverification requires a foundation of cosimulation to execute the design and codebugging to find the problem when things go wrong, but it's clear now that running software and debugging is not verification. This hierarchy of capability is shown in Figure 15.1.

15 System Verification

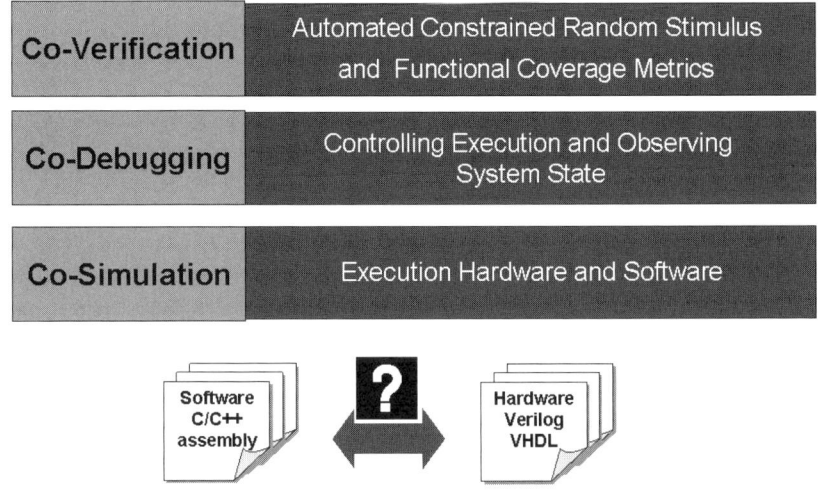

Figure 15.1 Layers of Capability

Advancing SoC Verification

To close the gaps in the three-stage verification process it is necessary to start to treat the embedded software more like the hardware. In the past embedded software has always taken a back seat in importance to hardware because software is "soft." As long as it can be changed and firmware updates and patches can be made available it is not treated with the same importance as hardware. Due to the quality issues previously discussed and the increased support costs it is becoming more important to treat the software like hardware, every corner case is important and every problem that can be found early leads to a big savings later. Consider some of the ways hardware verification is part of a rigorous process:

- Verification Planning
- Complex random generation
- Coverage points and metrics
- Tracking progress vs. a plan

For the purpose of SoC verification, the most important type of software is the low levels of software that interacts closely with the hardware. This is where all the hidden corner case problems exist. The application software is also important, but since it doesn't rely on the hardware details can rely on hardware abstraction and be developed outside the context of hardware verification.

There are three different kinds of corner cases to consider:

1. Corner cases exclusively in hardware.
2. Corner cases exclusively in software.
3. Corner cases that involve both hardware and software.

The type 3 corner cases are the ones that are most difficult to find and most often escape the three-step verification process. Finding type 3 corner cases holds the most potential for improving SoC verification.

Embedded software practices must advance beyond run and debug. One of the main reasons it's difficult to advance embedded software methods is the dependency on the hardware. With no hardware and now way to run software, the software engineers are limited in what they can do to improve the quality of software. As a result it's unlikely the software engineers wring the low-level code can drive improvements by themselves. The verification team is the glue that has the ability to advance the process and improve results. The next section discusses the challenges engineers face when they think about how to improve the three-step verification process.

List of Challenges

When the functional blocks are assembled to form the SoC there is a long list of challenges on the hardware verification front alone. Engineers that are well trained in verification would like to continue to use the same coverage-driven techniques they used at the block level to do generation, checking, and coverage. Figure 15.2 summarizes the environment the verification engineer would like to maintain. Besides the obvious problems of simulation performance,

15 System Verification

memory footprint, size of waveform dump files, etc. there is the all important question of what to do about the CPU and software. During block and subsystem verification engineers normally work without the CPU and use verification components to generate transactions on the CPU bus. The example presented in this paper uses an ARM926 CPU with a dual AHB interface. Verification engineers use the AHB eVC to perform coverage-driven verification and produce all of the corner cases on the bus, but the question remains about what to do at the SoC level.

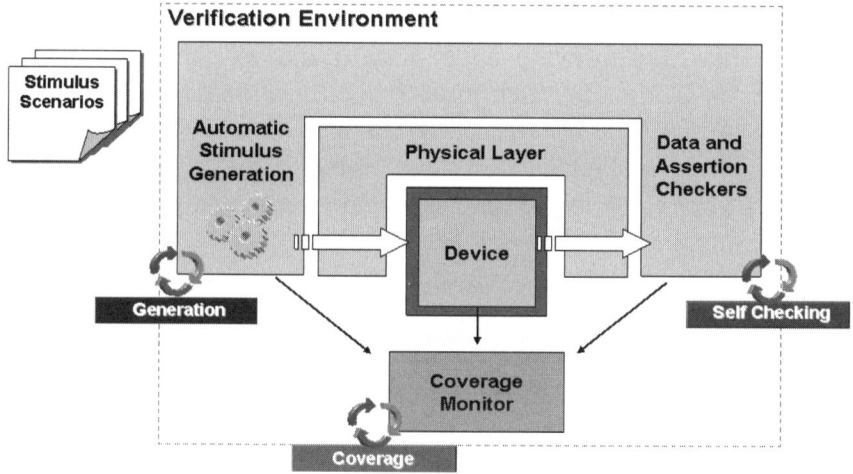

Figure 15.2 CDV Environment

There are two possibilities for the SoC simulation:
1. Continue to use the eVC
2. Use a full-functional CPU model

Working with the eVC makes it difficult to create activity that will be similar to the way the SoC will behave with CPU and software. For example, most eVC environments don't include details about interrupts such as executing instruction fetches from the interrupt vector when an interrupt occurs. Although constraints can be used to weight the transaction types, the transaction work load produced by the eVC is not similar to the CPU in terms of the address and data patterns. An even more important challenge is what to about SoC initialization. Some projects have reported thousands of programmable registers [2] that must be configured before the SoC is ready to do meaningful activity. Writing an initialization sequence for all of these registers takes a very long time. Besides being a tedious process, the motivation is somewhat low because in the end it's the job of the software to initialize the SoC, not the verification engineer. This leads to a duplication of effort.

The alternative to using an eVC is to insert the CPU and run software. Now the software can be used to initialize the SoC and the workload on the bus is much more realistic. The trouble with using software is to figure out what kind of software to use. There are many types of embedded software. Some software is created by engineers very close to the chip verification and some is created by other organizations in the company or other companies. Some of the common types of software include:

- System initialization software and hardware abstraction layer (HAL)
- Hardware diagnostic test suite
- Real-time operating system (RTOS)
- RTOS device drivers
- Application software

Certainly, initialization software is very useful for verification. On the surface the hardware diagnostic tests that are developed to run on the final hardware appear useful, but in the end are not as useful as originally thought for verification. Hardware diagnostics tend to be very simplistic and for the verification engineer skilled in CDV they tend to be too much like the directed Verilog testbenches used 10 years ago. They do something like: write, read, write, read, TEST PASS for each peripheral in the design. The RTOS and application software is typically not available yet and even if it is it would be too slow to run on anything except an emulator or FPGA board. The device drivers are interesting to run since they interact directly with the hardware, but for a driver to execute it needs to be called by an application (which is not available) so there is a stimulus problem that must be solved for the drivers to be useful.

In summary, there is currently a verification gap between the CDV approaches used by verification engineers to verify hardware and the embedded software, even at the lowest level of hardware diagnostics. These gaps result in projects using CDV for block and subsystem simulation, and then when they reach the SoC level make an abrupt change to run diagnostic C programs that do write, read, write, read, TEST PASS. This flow provides little or no chance to hit the important corner cases between hardware and software. The solution is to close the gap between CDV and the embedded software diagnostic program. Future sections will demonstrate how CDV can be used together with diagnostic software to provide more stress on the design and hit more corner cases at the SoC level.

ARM926 PrimeXsys Platform Design

To investigate the proposed solution to close the gap between CDV and diagnostic software an example SoC is used. The design is provided by ARM Ltd and is a reference SoC available from ARM as a starting point for a more complex SoC design. A block diagram of the design is shown in Figure 15.3. The design is about 500 k gates of logic plus memory. The CPU is the ARM926EJ-S which has a dual-AHB interface for instructions and data.

The PrimeXsys platform provides all the design data to use it in a larger SoC. Additionally, it includes verification components and software. The investigation for this paper centers on the DMA controller. The DMA controller is an ARM Primecell peripheral known as PL080. Most of the details of the DMA controller are not important, but below are a few highlights:

- 3 AHB interfaces
- 8 DMA Channels
- Memory to Memory, Memory to Peripheral, Peripheral to Memory, and Peripheral to Peripheral transfers
- Scatter gather DMA using linked lists
- Programmable burst length

- 8, 16, and 32-bit data transfers
- 66 Programmable registers

The platform is provided by ARM with three main verification environments as described in [3]:

1. *System Integration Environment*: No CPU model is instantiated. Instead, a BFM reads from files to generate traffic on the 2 CPU AHB interfaces. The goal is to verify connectivity of the peripherals and the memory map.
2. *System Verification Environment*: A CPU is instantiated and runs a set of diagnostic software programs to initialize the hardware test each peripheral.
3. *Scenario Validation Environment*: No CPU model is instantiated. Instead the AHB eVC is used. The goal is to hit corner cases and stress the design.

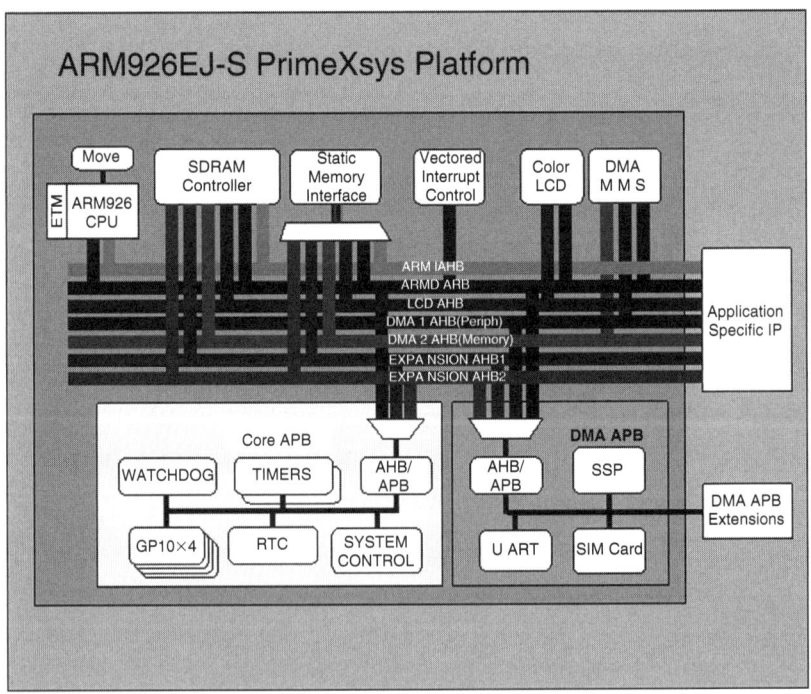

Figure 15.3 PrimeXsys Platform

Based on these three environments it's clear that number 1 provides the connectivity test to make sure the assembly is correct and there are no conflicts on the bus or in the memory map. Environment number 2 demonstrates the software diagnostics that are nice to run in verification and also later on FPGA boards or the final silicon. Although not stated, environment number 3 is likely a result of concern that the diagnostics are not comprehensive enough and the verification engineers understand the principles of CDV and feel like the environment number 3 to hit more corner cases.

These three environments are an excellent example of the verification gap between the software diagnostics (environment number 2) and CDV (environment number 3). The next section focuses on the solution to the gap by using the DMA controller as an example.

Closing the Gap

Coverage-driven verification requires a high level of control of the system. The only way to force the design into corner cases to meet coverage goals using constrained random generation is to monitor the state of the design and generate data that will move it to the required state to fill the coverage. A closer look at verification environment number 2 (diagnostic software) described in the previous section reveals that it suffers from a lack of visibility and coverage in the diagnostic software area. First, there is no way from the view of the software diagnostic to tell how well the test actually exercised the hardware. Second, there is no way to control the diagnostic program and coordinate it with the other hardware stimulus. Environment number 3 was developed because of these limitations. As also previously mentioned, environment number 3 has a different set of challenges since the SoC configuration must be done from an eVC (duplication of work) and the traffic is not as realistic as running a real software program on the CPU. The remainder of the paper will detail a new technique to unify environments 2 and 3 by utilizing the software diagnostic and CDV together. It will use the DMA controller and its diagnostic program as an example.

By reading the short description of the three verification environments it's clear there are limitations of what the hardware diagnostic programs can do and the engineers determined that to do an excellent job of verification they needed to perform CDV using environment 3. To their credit they did not stop with the first two environments and pass the design to software engineers as step 3 as has been described in the three-step verification process presented in the introduction. They saw a gap and tried to fill it with a CDV approach, but clearly now there are two separate environments to maintain and the situation is not ideal because of the lack of the CPU and software in the CDV environment.

DMA Diagnostic Program

Figure 15.4 is a code fragment of the DMA diagnostic used to verify the DMA controller functions correctly in the context of the SoC. This diagnostic is useful because it is a program running on the ARM926 and can easily be used again with hardware implementations of the design such as an FPGA board or the final silicon. The diagnostic is also good because it will demonstrate the mix of transactions

```
DMA_SetupData("DMA", (unsigned)&uTestData, 0x20000000, 0x10);     // Put source data in memory

nError += DMAC_M2M_Single_Transfer( (unsigned)(0x20000000),
                                    0x08002000, WORD_SI?           // Do single transfers

nError += DMAC_M2M_Single_Transfer( (unsigned)(0x20000004),
                                    0x08002000, HALFWORD_SIZE);

nError += DMAC_M2M_Single_Transfer( (unsigned)(0x2000000"
                                    0x08002000, BYTE_              // Do mutiple transfers

nError += DMAC_M2M_Multi_Transfer(  (unsigned)(0x20000000),
                                    0x08002000, 16      _SIZE);

nError += DMAC_M2M_Multi_Transfer(  (unsigned)(0x20000004),
                                    0x08002000, 16, HALFWORD_SIZE);

nError += DMAC_M2M_Multi_Transfer(  (unsigned)(0x2000000C),
                                    0x08002000, 16, BYTE_SIZE);

((uTestFail & DMA_FAIL) == DMA_FAIL) ?                             // Report Pass/Fail
    AvUtils_DEBUG_MSG("DMA: Tests Failed\n") :
    AvUtils_DEBUG_MSG("DMA: Tests Passed\n");
```

Figure 15.4 DMA Diagnostic Program

on the bus, the use of interrupts, and some possible effects of the caches on the hardware design. For a verification engineer there are clearly some drawbacks of the DMA diagnostic.

The first thing a verification engineer notices is the deterministic nature of the test. Verification engineers think about the interesting parameters of a DMA controller and where are the corner cases that need to be covered. The results of a short analysis of the DMA diagnostic are given in Table 15.1.

Table 15.1 Analysis of DMA Diagnostic

DMA source data	Fixed array: unsigned uTestData[] = {0x12345678, 0x55555555, 0xFEDCBA98, 0xAAAAAAAA, 0x89ABCDEF, 0x5555AAAA, 0x76543210, 0xABADCAFE, 0xDEADBEEF, 0xD0D0F00D, 0xA1C0FFEE, 0xCABBA6E5, 0xA6EDBEEF, 0x0DE2F00D, 0xD06F00D5, 0xF0E1D2C3} ;
DMA source address	Fixed at 0x20000000
Destination address	Fixed at 0x08002000
Bus widths used	8, 16, and 32-bits
Length of transfers	Single tests and always 16 for burst tests
Other programmable registers in the DMA controller	Always fixed with the same value
DMA controller modes used	Unknown, but complex modes of the DMA controller such as scatter/gather appear to always be disabled
Length of test	1 time through main() doing 6 DMA transfers

Clearly, it would be desirable to have a diagnostic program that would more fully exercise the DMA controller. Below are three options to improve the quality of this diagnostic program:

- Enhance the C program to be more comprehensive and cover every combination of interesting parameters including address and data values:
    o   Would probably take forever to write
    o   Is probably not necessary
- Take out the CPU and use an eVC to create a CDV environment to hit more corner cases:
    o   Doesn't leverage existing software (and software people)
    o   Utilizing a BFM results in different bus activity
- Utilize the existing C program and add the ability to call the C functions from a CDV environment, add randomization of the C functions called, the C data used, and add functional coverage to measure what the test really does:
    o   Very little extra work to create
    o   Utilizes existing software (and people)
    o   Uses the CPU in the system
    o   Uses the principles of CDV including coverage in both hardware and software

The last option is the way to close the gap in the three-step verification process described in the introduction. It's also the way to close the gap in the improved verification flow proposed by ARM in the PrimeXsys platform that used two separate verification environments, one for diagnostics and one for CDV. By connecting and controlling the C functions to the verification environment the best of both world's can be achieved, running C functions and the generation, checking, and coverage provided by CDV.

The Generic Software Adapter
The Generic Software Adapter is a Specman adapter used to connect to and control embedded software. Most engineers know that Specman contains ways to connect verification environments written in *e* to designs under verification (DUV) written in Verilog, VHDL, and SystemC. GSA connects the *e* environment to embedded

software, such as the DMA diagnostic from the PrimeXsys platform. Now the verification environment can control both the hardware design and the software running on the ARM926 CPU. GSA works in any cosimulation environment using any type of CPU model. Some example environments it has been used with are:

- Verilog or VHDL RTL CPU model running in a logic simulator
- Instruction Set simulator connected to an HDL BFM in a logic simulator
- Host-code execution where software is run on the host machine and connected to an *e* BFM using the Specman coverification link (CVL) [4]
- SystemC TLM simulation of an SoC
- Emulation Systems such as Xtreme and Palladium:
    o RTL CPU models inside the emulators
    o In-circuit emulation using a CPU board connected to the emulator such as the ARM Logic Tile

GSA uses a shared mailbox memory that is located somewhere in the memory map already defined and modeled in the SoC. The embedded software can communicate with the memory by software instructions and the verification environment can communicate with the mailbox memory using the backdoor techniques such as the Verilog variable statement already available in *e* or other suitable interfaces based on the type of memory model used (such as Denali). For details of the communication mechanisms and more examples refer to [5].

To connect to the DMA diagnostic C functions and variables to the verification environment *e* ports are used. The *e* language has different types of ports available to communicate with the DUV (the embedded software in this case). GSA uses method ports to call C functions and simple and event ports to read and write variables in the C code.

The key to GSA is the unique ability of Specman to "generate stubs." The stub generation process enables the connection between the verification environment and the embedded software to be completely automated. GSA finds all the ports in the *e* environment and automatically generates the C code to provide the communication and the user can simply link this automatically generated C code with the DMA diagnostic. The overall process to perform CDV with embedded software is shown in Figure 15.5. The next section shows how this was done for the PrimeXsys DMA diagnostic.

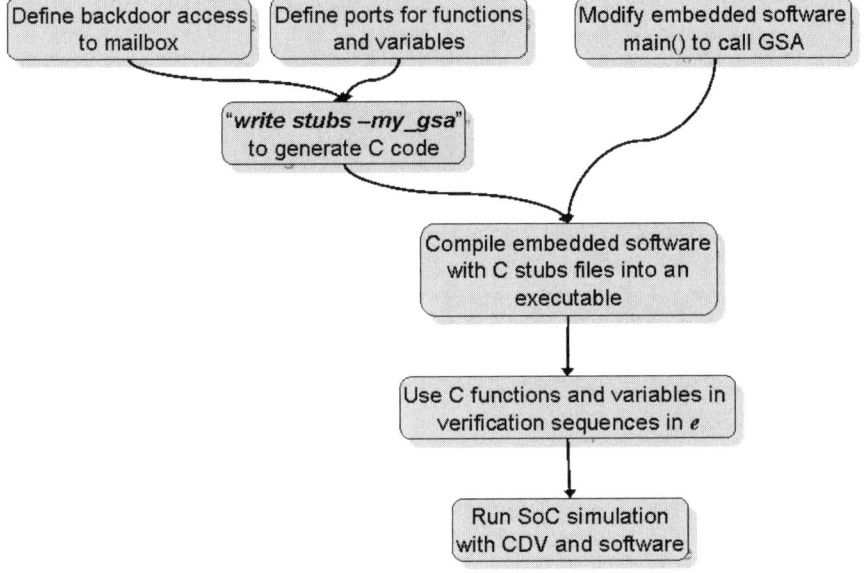

Figure 15.5 GSA Integration Flow

Connecting the DMA Diagnostic to the Verification Environment

To improve the verification quality of the DMA diagnostic program the Generic Software Adapter was connected to the existing DMA C functions. The goal is to improve verification by exposing more corner cases using the principles of coverage-driven verification as compared to the existing C test which is completely deterministic. The basic steps required to connect the environment to the embedded software are reviewed in the next section.

Memory Connection

The first step in GSA integration is to enable the verification environment to access the memory that will be used for the communication mailbox. This is done by creating backdoor memory access functions that can read and write memory without advancing simulation time. GSA defines an interface for the verification environment designer, called a mini-adapter, to connect the appropriate memory model being used as the mailbox. Because each design has a different memory map and uses different types of memory models this step requires some manual coding. In the case of the PrimeXsys platform the tightly coupled data memory (DTCM) was used as the mailbox memory. The ARM926 includes dedicated interfaces to fast memory called TCM that is directly connected to the processor. Figure 15.6 shows a fragment of code used to implement the mini-adapter by connecting to the DTCM for accesses from the verification environment. The DTCM is modeled using a Verilog memory array and the *e* Verilog variable statement is used to provide easy access from *e*. It's important to note that since the DTCM is a local instance of memory the mini-adapter must have some knowledge about the address in the ARM CPU memory map where the DTCM memory resides. The mini-adapter uses this information to compute the correct addresses of the DTCM memory instance.

```
unit vr_pwp_verilog_if {

  mem_base: uint;

  verilog variable
      'TBplatform.uPlatform.uProcSubSys.uProcCoreMod.uDRAM.
                                          ram.memory[8191:0][31:0]';

  read_int(a: uint): uint is {
     var address: uint;
     address = a - mem_base;
     result =
     'TBplatform.uPlatform.uProcSubSys.uProcCoreMod.uDRAM.
                                          ram.memory[address[31:2]]';
  };

  write_int(a: uint, data: uint) is {
     var address: uint;
     address = a - mem_base;
     'TBplatform.uPlatform.uProcSubSys.uProcCoreMod.uDRAM.
                                          ram.memory[address[31:2]]'= data;
  };
};
```

Figure 15.6 Mini-adapter Code

Port Definitions

Once the mailbox memory is connected, the next step in GSA setup is to define the ports that will be used. This includes the method ports that will be used to call the C functions in the DMA diagnostic and the simple ports that will be used in the verification environment. The creation of ports can be automated either by the use of an *e* macro or by using the Verification Builder, a GUI tool for environment creation, but for the purposes of showing how GSA connects the source code of the port definitions will be shown. For simplicity only a small part of the DMA diagnostic is shown.

The first area of interest is the data used in the DMA. Recall that in the original DMA diagnostic this data consisted of a fixed array of data as shown in the first row of Table 15.1. To improve verification new random data should be used for each DMA transfer. To randomize the data a simple port is created in the verification environment that will connect to the C variable uTestData. The simple port is of direction inout so it can read the C data array and also write it with new random data.

Recall from Figure 15.4 that one of the C functions in the DMA diagnostic was DMAC_M2M_Multi_Transfer(). This C function does a multiple word transfer between two memory locations. The destination memory address, the number of words, and the width (8, 16, or 32 bits) are all arguments to this function. In the original DMA test all of these arguments are fixed. To improve verification this function should be called more than just three times and with random arguments. To do this an out method port is created in the *e* environment. A fragment of the port definitions used is shown in Figure 15.7.

Not all of the ports are shown, but additional ports were created for all of the C variables and functions that are accessed from the verification environment.

15 System Verification

```
method_type DMAC_M2M_Multi_Transfer(src_addr: uint, dest: uint,
                            num: uint, width: uint): uint @sys.any;

unit vr_pwp_env like any_env {

   logger : message_logger is instance;
   name : vr_pwp_name;

   short_name(): string is {
      result = append(name);
   };
   short_name_style(): vt_style is {
      result = ORANGE;
   };
   show_banner() is also {
      out("(c) Cadence 2006");
      out("vr_pwp instance : ", name);
   };
   show_status() is only {
      out("vr_pwp Verification Environment - instance : ", name);
   };

   // simple port for generating random data
   uTestData: inout simple_port of list of uint is instance;
   keep bind(uTestData, external);
   keep uTestData.hdl_path() == "uTestData";
   keep uTestData.external_type() == "unsigned int [32]";

   // C function to do multi-word DMA
   DMAC_M2M_Multi_Transfer: out method_port of
                           DMAC_M2M_Multi_Transfer is instance;
   keep bind(DMAC_M2M_Multi_Transfer, external);
   keep DMAC_M2M_Multi_Transfer.hdl_path() ==
                                 "DMAC_M2M_Multi_Transfer";
```

Figure 15.7 Port Definitions

Connecting the Main() Function in C

The original DMA diagnostic defined a main() function and proceeded to call the DMA C functions. Recall, it called the two DMA functions available for single and multiword transfers a few times each. GSA takes care of all the calling of the C functions as specified by the *e* environment. Depending on the test developed this could mean calling the functions many times or just a few. To improve verification sequences are used to setup interesting corner cases to stress the DMA controller in the context of the SoC. To facilitate the use of sequences the main() function will simply yield to the verification environment and receive commands by the underlying mailbox architecture. The modified main() function is shown in Figure 15.8.

```
int main(void)
{
   // Initialize GSA
   sn_gsa_init();

   // no need to modify the messages already in place
   AvUtils_DEBUG_MSG("platform_DMA_test: Starting test\n") ;

   // loop forever processing C calls from the e environment
   while (1) {
       sn_gsa_wait();
   }
}
```

Figure 15.8 Modified main Function

The modified main() function has replaced the directed set of C function calls with a loop that will receive much more random calls from the e environment. The other parts of the main() function such as #include files and global variables were not modified.

Writing Stubs

Once the environment is complete, the next step is to create the "stubs" files. Specman users will be familiar with the stubs file for languages such as Verilog and VHDL. The concept for GSA is the same, but the stubs file written is in C. The automatically generated stubs file takes care of the underlying mailbox protocol to make GSA possible. Once generated, the C stubs file is compiled with the DMA diagnostic and linked into the executable. Functions such as sn_gsa_init() and sn_gsa_wait() as shown in Figure 15.8 are part of the stubs file. Below is the command to write the stubs file for GSA (C file) and for NC-Verilog (Verilog file).

```
% specman -c "load vr_pwp/examples/test1;
write stub -gsa_pwp ./vcode/gsa_pwp_specman;
write stub -ncvlog"
```

Notice that GSA completely automates all of the method port and simple port connection using this automatic stubs generation.

Creating Sequences and Coverage

The final step to put everything together is to create interesting sequences to call the DMA C functions with random data and arguments and collect coverage on interesting activity. Multiple sequences were created for the PrimeXsys platform and different coverage values were collected.

```
extend MULTI_TRANSFER sw_sequence_item {
   errors: uint;
   src_addr: uint;
   dest: uint;
   keep soft dest in [0x08002000..0x09000000];
   keep dest[1:0] == 0;
   num: uint;
   width: vr_pwp_width;
   keep soft num in [1..16];

   activate() @driver.clock is {
      --method_type DMAC_M2M_Multi_Transfer(src_addr: uint,
                   dest: uint, num: uint, width: uint) uint @sys.any;
      errors = driver.p_env.DMAC_M2M_Multi_Transfer$(src_addr,
                                      dest, num, width.as_a(uint));
   };
   nice_string(): string is also {
      result = "DMAC_M2M_Multi_Transfer()";
   };
};

extend ST2 sw_sequence{
   src: uint;
   keep soft src in [0x20000000..0x20001000];
   keep src[1:0] == 0;

   !setup: SETUP sw_sequence_item;
   keep setup.uDestAddr == src;

   !multi: MULTI_TRANSFER sw_sequence_item;
   !errors: uint;
   keep multi.src_addr == src;

   body() @driver.clock is only {
      do setup; // copy data to source address
      do multi; // do single DMA
      errors = multi.errors;
   };
   nice_string(): string is also {
      result = "ST2";
   };
};
```

Figure 15.9 Example Sequence

For simplicity, just one sequence is shown to create a multiword DMA transfer. Figure 15.9 shows the sequence item created with random arguments for the C function and constraints to keep the random arguments within the memory map. This is followed by creation of the sequence ST2 to first setup the DMA data to be used and then call the multiword DMA sequence item to transfer the data.

The complete verification environment for the DMA diagnostic was created to create multiple sequences of interesting scenarios calling the DMA C functions with random and collecting coverage on the software and combining the software coverage with the hardware coverage. The last step is to create a test that runs a mix of sequences. One of the tests is shown in Figure 15.10. This test creates and runs 150 sequences using ST1 for half of them and ST2 for the other half. Given the sequence library the test writer can choose any mix of sequences to perform verification and can also run with a different random seed to create a unique stimulus.

```
<'
import vr_pwp/examples/vr_pwp_config.e;

extend MAIN sw_sequence {
   keep count == 150;
   keep sequence.kind == select {
      50: ST1;
      50: ST2;
   };
};

'>
```

Figure 15.10 Example Test

Results

After integrating CDV with the DMA diagnostic a new analysis was done of the results of the verification environment. Augmenting the existing C DMA diagnostic functions with GSA has brought the concepts of generation and functional coverage to the embedded software. The result is 1000s of DMA transfers with randomized parameters instead of the directed test of six DMA tests and the pass/fail message (Table 15.2).

Table 15.2 Analysis of DMA Diagnostic with GSA and CDV

DMA source data	Randomized
DMA source address	Randomized within ranges of memory map
Destination address	Randomized within ranges of memory map
Bus widths used	Randomly generated 8, 16, and 32-bits
Length of transfers	Single tests and randomly generated lengths
Other programmable registers in the DMA controller	Multiple values constrained by random generation
DMA controller modes used	Measured by functional coverage
Length of test	Controlled by *e* environment, 1000s of DMAs with different sequences can easily be run

The DMA example can be extended to the complete suite of diagnostic software available with the PrimeXsys platform and many tests can be run on in parallel to create interesting corner cases with little effort by the verification or software teams.

Conclusion

The use of GSA has been proposed and demonstrated to be a solution to the commonly used three-step SoC verification process that has difficulty catching the corner case problems between hardware and software.

- Perform comprehensive verification of each functional block of the design
- Assemble the blocks together to form the SoC
- Execute Software on the SoC

In the past, these corner case escapes were caused by the inability to control the embedded system software. The ARM PrimeXsys platform was examined and it was shown that the verification team did recognize the gap between C diagnostics and CDV, but were forced to create two separate environments to solve it, one for diagnostics and one for CDV. This paper has demonstrated how to unify the existing C diagnostics with CDV and greatly increase the ability to hit corner cases while still utilizing the existing set of C diagnostics.

References

1. Co-Verification of Hardware and Software for ARM SoC Design by Jason Andrews (http://coverification.home.comcast.net).
2. "Extending a coverage driven verification environment with real software" by Ernst Zwingenberger, Micronas, CDNLive! EMEA, Nice, France, June 25–27, 2006.
3. ARM PrimeXsys Virtual Component Verification Environment Reference Manual, ARM Ltd., 2004.
4. Specman Usage and Concepts Guide for e Testbenches, Version 5.0.3, Chapter 12, Using the Co-Verification Link (CVL).
5. Hardware Software co-verification using Coverage Driven Verification Techniques, Giles Hall, Cadence Design Systems, 2005.

Chapter 16
Mixed Analog and Digital Verification

Coverage-Driven Verification for Mixed-Signal Circuits

Monia Chiavacci, Egidio Pescari and Gabriele Zarri
Yogitech

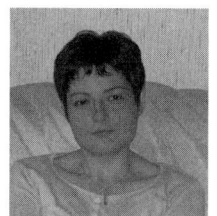

Monia Chiavacci

Ms. Chiavacci cofounded Yogitech in 2000. She is responsible for the mixed-signal division. She worked as an analog designer from 1998 to 2000 after receiving her degree cum laude in Electronic Engineering at the Pisa University. Her work experiences include high-reliability systems in critical environments such as biomedical, space and high-voltage automotive applications.

Gabriele Zarri

Mr. Zarri is a verification engineer at Yogitech. He is responsible for the development of verification IPs, verification environments for many international customers, and trainings on verification methodologies. His experience includes automotive protocols such as LIN, CAN, and Flexray. He is expert in OCP protocol, a universal complete socket standard for SoC design, and has recently acquired experience in the verification of mixed-signal circuits. Gabriele specialized in Microelectronics and Telecommunications with a MS from Nice Sophia-Antipolis University.

Egidio Pescari

Egidio is a senior design & verification engineer at Yogitech. Prior to Yogitech, Mr. Pescari developed systems in critical environments such as automotive and space applications. He acquired experience in many automotive protocols such as LIN and CAN. He graduated from the University of Perugia in 1998.

Abstract

Traditional methodologies for Analog and Mixed-Signal (AMS) verification present many drawbacks.

Analog design verification is usually subjective due to the lack of automatic checks and the poor control on stimuli and results. Moreover, verification of mixed-mode circuits is often incomplete due to fact that analog and digital macros are simulated with two different environments with insufficient interaction.

Measuring the quality of verification becomes difficult, costs escalate in redesign, engineer-time and market entry is unpredictable. Moreover, lack of reuse in verification environments results in lower levels of efficiency.

This paper describes an innovative Analog Mixed-Signal Verification methodology based on a coverage-driven approach which extends to the analog domain well-know concepts in the digital one, achieving advantages in terms of completeness, effectiveness, process control, and reusability. An introduction is given on the basic items on which the verification methodology is built, how to define a verification plan including analog metrics for functional coverage evaluation is described together with a tool bridging the analog and digital domains.

Introduction

The growth of the monolithic mixed-signal systems foreseen for the near future drives EDA vendors and SoC solution providers to invest huge resources to explore new verification approaches covering the gap between the analog and the digital verification current status. Even if mixed-signal simulator tools are already available in the market linking and simulating analog and digital blocks in the same test-bench, it is still not possible to extend to analog mixed-signal domain the advanced techniques already available for digital verification, such as constraints capture, randomized or pseudo-randomized stimulus generation and self-checking results collection with coverage analysis.

Thus, most analog designers simulate various mixed subsystems in order to verify different functionalities of the whole device (i.e., reset and start-up conditions, power down/up signals polarity, functionalities of analog block configured by a digital block, test-mode, etc.): the space of the interactions among all subcells that may be derived from the device specifications is most of the time not completely covered so that verification coverage is often partial and even not quantified. Lack in methodology and automation amplifies the risk of subjective evaluation and reduces reusability. In this paper, starting from previous work [1–3], we are presenting an innovative analog and mixed-signal verification approach in which both analog stimuli and output metrics can be generated in an advanced digital verification environment.

Traditional Mixed-Signal Verification

Analog- and digital-design processes are fully separated: different teams, different expertise, different tool chains, and often different cultures. Finally analog and digital sections must work together on silicon in a mixed-signal circuit and so becomes mandatory and more and more urgent to set up a mixed-signal approach shared between the two sides of the same house.

Currently functional verification in analog domain has a lot of limitations compared to the digital one. Mixed-signal verification comes from the analog side and has the same amount of problems or even more due to the multidomain nature. As shown in Figure 16.1, AMS verification is at least two steps behind the digital one and is moving forward slowly.

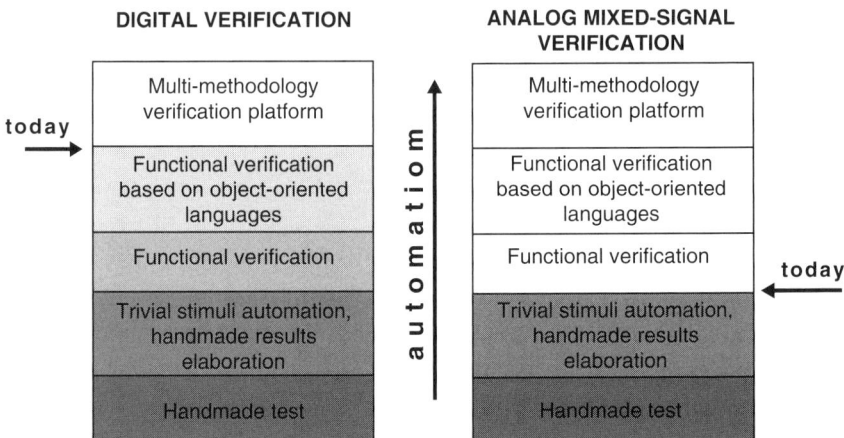

Figure 16.1 Comparison between Digital and Analog Mixed-signal Verification approach

As consequence of the lack of a proven methodology and the low control of the overall verification process, mix-mode simulations handling in the same test-bench analog and digital blocks are mainly driven by engineer's experience on the specific design or application. Such approach produces a strong limit in resource flexibility and optimization. The low level of reuse is another negative consequence which increases the effort needed at each design cycle.

In a traditional approach, mixed-signal verification is often in charge of the analog team and based on direct tests: the analog designer usually creates and fills a spreadsheet listing all the needed tests and then removing part of them based on his experience on circuit implementation. This procedure limits, by nature, the reuse level for other circuits and increases the risk to overlook or miss something. At this point the designer creates test-benches covering the listed tests, performs simulations, checks results mainly by visual inspection and fills the spread sheet with measurements results. In this approach the huge lack in automation forces high-value resources to take care of time-consuming and repetitive tasks.

The increase of complexity and the increase of costs, both in masks for deep-submicron technology and in time-to-market mismatch, drive the requirements of powerful methodologies and tools able to solve those issues.

Verification Planning

The coverage-driven verification methodology can satisfy the requirements also for mixed-signal circuits.

The verification plan is the key element for the circuit verification and it enables the coverage-driven approach. It must be defined at the very beginning and it must describe all the requirements and results for the verification activity collecting and organizing the contribution of the different expertise necessary to make the design successful: analog design team, digital design team, and verification team.

A coverage-driven verification approach uses predefined metrics to evaluate the verification progress, i.e., to measure the amount of covered conditions according to the defined metrics respect to the complete set. Whatever is the mechanism to create test-cases, the functional coverage measures the percentage of the test space covered by the test-cases run at a certain point of the verification process.

How to define and measure metrics in a digital context is well know [4] and the extension to the analog one could be quite straightforward starting from what is available in the discrete domain. To allow the adoption of this approach in an actual project, it is necessary to provide a tool managing analog metrics for functional coverage definition.

Let's consider a very simple example to better understand how the verification plan can be created for an analog circuit and how to define the methodology extending to the analog and mixed-signal circuits the coverage-driven functional verification.

A peak detector is shown in Figure 16.2 with the input and output waveform and the transfer function.

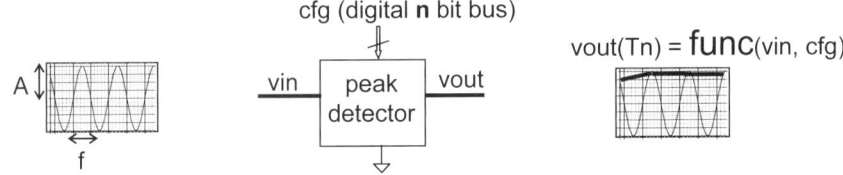

Figure 16.2 Peak Detector with Sinusoidal Waveform in Input, Digital Configuration Word Defining Output Parameter as Gain and Offset and Output Waveform According to its Transfer Function

The input of a verification plan is the specification document and it is mainly composed by four sections as described in the following.

1. *Definition of the primary input space and the set of device states.*
   In this simple example the input space is mainly composed by the parameters of the input sinusoidal waveform (A and f) and some environment conditions as temperature and technology process model cards.

   $$A \in [A_{MIN}, A_{MAX}]$$
   $$f \in [freq_{MIN}, freq_{MAX}] \quad (1)$$

   $$T \in [T_{MIN}, T_{MAX}]$$
   model card= mod_0, mod_1, mod_2…..mod_n

   Instead, device states are related to configurations depending on the digital input bus. In this simple case the bus is a primary input but in a more complex circuit it could be an internal signal.

   $$cfg \in [0, ..., k] \quad (2)$$

   Parameters and their values come from the specification documents.

2. *Definition of verification items*, i.e., the list of device functionalities to be verified according to the specification document.

16 Mixed Analog and Digital Verification

For each items the plan defines:
1. Item ID
2. Functionality description
3. Conditions at which the item has to be verified (they depend on stimuli, on external environment parameters and on device states)
4. Description of measurement procedure (i.e., how to extract the functionality)
5. Expected results

In our simple case, we can consider only one item.

Output voltage value	For each possible value of input amplitude and frequency, in each possible environment conditions and configurations	Sample output voltage value when the input signal reaches the maximum voltage value within one period	vout(Tn) = func(vin, cfg)

3. *Definition of metrics to measure functional coverage.*
   In this section the attention has to be put in the definition of the rules to be used to measure the functional coverage. Ideally, the circuit has to be verified in every condition defined in the first step.
   There are mainly three elements to be considered in order to limit the space of the metrics and so the total amount of test cases.

   a. For a given parameters, not all the possible values are allowed.
      For instance, for the configuration bus at n bits the allowed values are only "k+1" according to the specification reported in (2)
   b. In analog domain there are continuous quantities (e.g., voltage values in a defined range) but contiguous values "normally" do not create distinguishing behavior of the circuit.
      For instanced, applying $A = 2.1$ V after $A = 2$ V do not add any information for the verification of the

circuit; instead, if for the nmos device $Vt = 0.65$ V, depending on circuit implementation, for $A = 0.6$ V and $A = 0.7$ V the circuit can behaves in a different way.

So, for each continuous value parameter, it is possible to define subranges where, for value in the same subrange the behavior is expected to be the same, instead two values in two different ranges can produce quite different response. Based on that makes no sense to test more than one value in each range but it is important to test at least one value for each subrange.

c. For some continuous quantities, based on experience, circuit implementation and so on, it is clear that only the boundaries of the defined range are critical for the circuit behavior.

It is the case, for instance, for the environment temperature: normally only the minimum and the maximum temperature according to the specification are considered in corner analysis.

Input items:

ID	Name	Buckets	Description
#1	AMP	$[A_{MIN}, A_1]$, $(A_1, A_2]$ ... $(A_h, A_{MAX}]$	Input amplitude
#2	FREQ	$[f_{MIN}, f_1]$, $(f_1, f_2]$, ... $(f_k, f_{MAX}]$	Input frequency

States items:

ID	Name	Buckets	Description
#3	CONF	[0], [1], ... [k]	Conf. word

Output items:

ID	Name	Buckets	Description
#4	VOUT	$[V_{MIN}, V_1]$, $(V_1, V_2]$ ... $(V_j, V_{MAX}]$	Ouput voltage

Cross-coverage

AMP&FREQ&VOUT
CONF&FREQ&VOUT

Figure 16.3 Build up Coverage Items and Cross-Coverage Ones

Coverage items are related to primary input space, device space and output space.

Metrics can be then created by the combination of coverage items. For instance, in our simple case we can define the items as reported in Figure 16.3.

4. The last part of the verification plan has to describe the verification scenarios needed to check all the listed items in (2) and reach the target functional coverage defined in (3).
   In fact, especially in the analog domain, it is not always possible or optimum to perform all the checks with the same test-bench, with the same topology. For some verification items could be necessary to apply some stimuli and the generation mechanism could be incompatible with the one needed for others items. For instances the circuit should be normally powered with a DC generator to check its functionality. In case of start up test, the power supply voltage has to be provided by something like a pulse generator. In this case to reach stationary conditions can require some time (the transient behavior at start up has to end before to do functional test). It's more convenient then to have two different topologies: one for start up scenario and one for functional scenario.
   These test scenarios have to be identified and described in the verification plan in order to give defined guidelines for the verification environment implementation.

Once the verification is ready the verification environment can be implemented. It is important to highlight that the methodology shortly described in this section helps to maintain an update and fully consistence specification document for the mixed-signal circuit: this helps to solve another weak point in mixed-signal design.

Analog Mixed-Signal Verification Kit

To introduce the proposed verification methodology for mixed-signal circuit it is necessary to provide a tool able to support such methodology and providing enough automation in order to avoid engineer-time to be spent checking its correct usage.

The Yogitech Analog Mixed-Signal Verification Kit (AMSvKit) is a tool able to link analog and digital approach; AMSvKit extends to analog domain verification techniques already used in digital one providing a unified environment for mixed-signal verification, based on Cadence Specman Elite [4, 5].

As illustrated in Figure 16.4 the analog mixed-signal verification kit is composed of three libraries (*vTerminals, vComponents* and *Sequences DB*) and all the necessary infrastructure to make working the full environment (simulator scripts, Specman "*e*" language structures/unit, etc.).

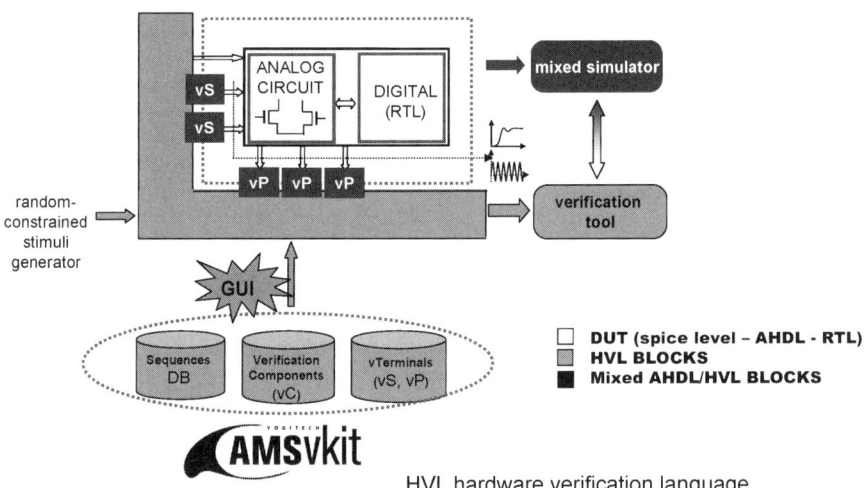

Figure 16.4 AMS vKit Scenario

The core of the kit is a library of "verification terminals" (*vTerminals*) that creates an interface between the analog and digital domains. The vTerminals are divided into two types:

- Verification sources (*vSources – vS*), which are models of signal sources configured and controlled by digital commands from the verification environment that provide continuous and time-continuous voltage and current signals or analog events; they

include DC, pulse and sinusoidal signal (current and voltage) generators, noise injectors, and parameter spread emulators;
- Verification probes (*vProbes* – *vP*), which transfer analog information from the mixed-signal simulator to the verification environment; they provide values of voltage, current, and timing parameters and include self-checking mechanisms (e.g., check a sampled voltage level within a predefined range); examples of vProbes are voltage/current/time detectors, linear behavior and total harmonic distortion calculators, AC gain extractor, etc.

The verification components (*vComponents*) are ready-to-use to create verification environments (e.g., test-benches) for main blocks, including self-checking mechanisms and coverage evaluation based on analog metrics that are easy to integrate in more complex mixed-signal scenarios. They are developed to verify basic analog blocks such as band gap cells, oscillators, voltage regulators, comparators, operational amplifiers, and buffers.

For each cell, the verification plan has been defined including the significant parameters, conditions and procedures to measure them. Based on the verification plan, the verification component drives, monitors and processes current and voltage signals generating correct stimuli for the DUT and elaborates the information in order to match the target coverage.

In order to calculate a nontrivial analog parameter it is necessary to properly control and configure a number of vSources and vProbes and to synchronize them. This is implemented using sequences: a structure that represents a stream of items signifying a high-level scenario of stimuli.

The database provided with the kit (*sequences DB*) includes all the sequences needed in an analog context. For instance, in order to extract the total harmonic distortion of a buffer (one of the most important analog parameters), it is necessary to stimulate the circuit with a sinusoidal signal (vSource) for a defined time period depending on the frequency at which the measurement has to be done. The settling time and the sample period of the output signal (vProbe) depend on the frequency as well. Predefined and ready to use sequences which create this and other kinds of test scenarios are available in the sequence DB library.

Using AMS vKit the powerful generators of state of the art verification tools such as Cadence Specman Elite [4, 5] can be used to generate also analog stimuli; checking mechanisms can be applied to analog verification items and functional coverage can be evaluated also based on analog metrics according to the defined verification plan.

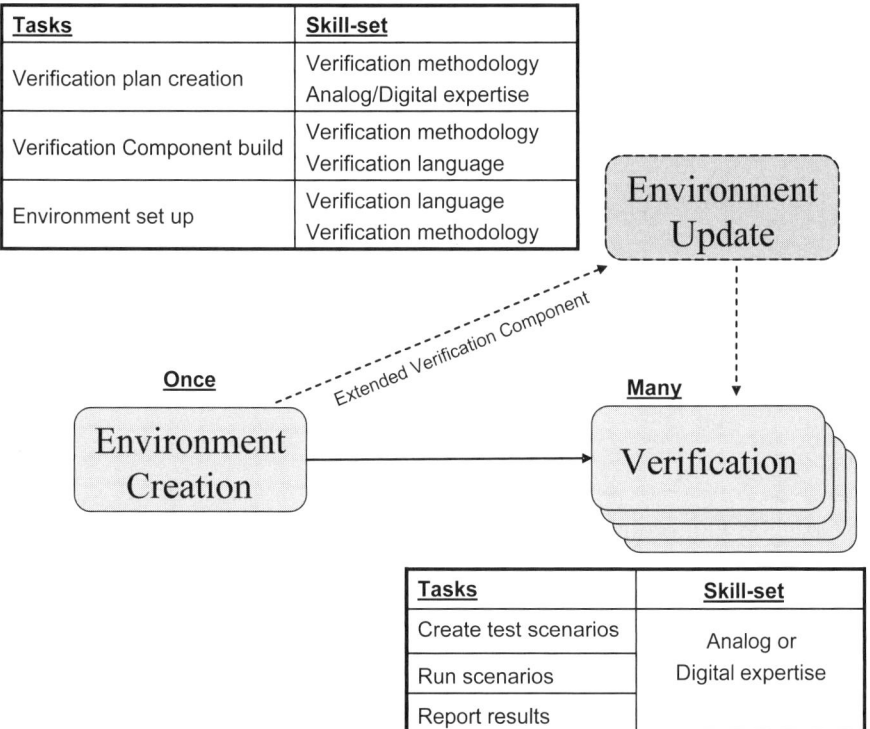

Figure 16.5 Skills Set for the Verification Flow

Figure 16.5 describes stages and skills set for the verification process. The environment creation starting from the verification plan definition is a high-value step for which the complete set of expertise is needed: analog, digital and verification.

For the verification itself, i.e., simulation run, both digital and analog designer are able to perform the task giving in this way a high level of flexibility in term of resource allocation: this is a big advantage on top of the whole methodology.

Coming back to the simple example, the AMS vKit provides:

- vSources to generate analog stimuli
- vProbes to monitor and process analog information
- Capability to handle parameters, generating random-constrained analog values
- Capability to implement self-checking mechanism for analog items
- Capability to handle analog metrics for functional coverage evaluation

Conclusion

In summary, the AMSvKit allows the extension and the adaptation of the three main steps for a powerful coverage-driven functional verification to mixed-signal circuits verification:

1. Random-constrained stimuli generation
2. Self-checking mechanisms
3. Functional coverage evaluation

This enables and deploys the described methodology to introduce automation in mixed-signal verification, extending the coverage-driven approach to analog/ mixed and increasing verification quality, effectiveness and reusability.

Flexibility in resource allocation due to the methodology and the automation is another important added value of the described approach: traditionally mixed-signal verification is performed by analog designers taking away them from pure design tasks for which experience and creativity is mandatory.

The introduction of a methodology in verification forces development team to pay more attention to critical element like specification document: this is an input of the verification and it must be clear, complete and update. Often this is not true and the consequence can be dramatic.

Moreover, having a supervisor as the verification tool together with coverage metrics allows tracing back the contribution of each simulation to the functional coverage. So, it is also possible to optimize the run time selecting the simulations which give higher contribution in terms of coverage respect to the defined verification plan. Automatic verification management tools giving such control for digital verification can be used also in mixed-signal context.

Reference

1. R. Mariani, M. Chiavacci, G. Bonfini "Fundamentals of a novel approach for mixed analog-digital verification", *9th IEEE European Test Symposium*, Informal Session, Ajaccio (Corsica), 23–26 May 2004
2. G. Bonfini, M. Chiavacci, R. Mariani, R. Saletti "A new verification approach for mixed-signal systems", *2005 IEEE International Behavioral Modeling and Simulation Conference(BMAS 2005)*, 22–23 September 2005, San Jose, California, USA, accepted for web publication
3. G. Bonfini, M. Chiavacci, F. Colucci, F. Gronchi, R. Mariani, E. Pescari, A. Sterpin "Fault coverage in a new mixed-signal verification environment", *In Proc. of 11th International Mixed-Signal Testing Workshop (IMSTW)*, 27–29 June 2005, Cannes, France, pp. 148–154
4. Cadence's Specman tool, www.cadence.com
5. IEEE 1647: http://www.ieee1647.org/index.html

Chapter 17
Design for Test

Over the past decade, several advances in structured testing for manufacturability and reliability have contributed to the automation of verification processes. In this chapter by Stylianos Diamantidis, Iraklis Diamantidis, and Thanasis Oikonomou of GlobeTech Solutions, we will see how verification technologies can be used to create a complete, fully automated unified solution from test specification to DFT closure.

Stylianos Diamantidis

Stylianos Diamantidis is a founder of Globetech Solutions, where he currently serves as Managing Director and CTO. With over 10 years of experience in design verification, he is responsible for driving IP product strategy, engineering, and consulting services. Prior to Globetech, Stylianos managed system-level diagnostic software development at Silicon Graphics Inc., spanning across server, supercomputer, and graphics products. Stylianos holds a B.Eng. in Computer Systems Engineering from the University of Kent at Canterbury, UK, and an MS in Electrical Engineering from Stanford University.

Iraklis Diamantidis

Iraklis Diamantidis is a founder and Senior Verification Engineer at Globetech Solutions. His current areas of interest include electronic system-level design, advanced design verification methodologies, silicon test, debug and diagnosis, and system software. Iraklis holds a B.Eng. from the University of Kent at Canterbury, UK, and an MS in Electrical Engineering from Stanford University. He is a member of the IET and the IEEE.

Thanasis Oikonomou

Thanasis Oikonomou is a Senior Digital Systems Designer and Verification Engineer at Globetech Solutions. His interests include computer architecture, high-speed networks, digital design, verification, and testing. He received B.Sc. and M.Sc. in Computer Science from the University of Crete, Greece.

A Unified DFT Verification Methodology

Stylianos Diamantidis, Iraklis Diamantidis, and Thanasis Oikonomou
Globetech Solutions
Thessaloniki Greece

In today's fast growing SoC, incomplete or ineffective DFT support due to poor specification or loose design practices can quickly become the critical path to making market windows and delivering products within cost restrictions.

This paper will introduce a unified DFT Verification Methodology (DFT-VM), aimed at providing a complete, methodical, and fully automated path from test specification to DFT closure. We will also examine the benefits of this approach, looking at how this methodology can help bridge the widening gap between design and test.

Introduction

As modern IC transistor counts continue their frenzied climb according to Moore's Law, test infrastructures, the collection of logic dedicated to testing the structural integrity of silicon are also fast growing in both area and complexity.[1] In a nanometer design era where silicon debug already takes up to 30% of project time and

---

[1] In recent studies, DFT in ICs has been found to typically account for 20% of total logic gates and for as much as 30% of total design effort.

semiconductor test cost typically accounts for 30–50% of total fabrication cost, Design-For-Test, or DFT, is assuming a critical role in product definition, design, and delivery.

Although DFT is a concept that has been around for a long time, semiconductor companies are today experiencing unprecedented pressure to provide more complex DFT features in their designs. This trend is largely attributed to the need for controllability and observability within highly integrated SoCs and is driven by the inevitabilities of test economics.

Design verification tools and methodologies have made tremendous progress in the last few years, directly benefiting design quality and shortening development cycles. However, DFT-specific circuitry tends to be overlooked in most test plans. There have been a series of contributing factors for this oversight:

- No clean test intent is specified and communicated to the design teams
- Lack of formal end-goal or associated Quality of Result (QoR) for DFT
- Low prioritization compared to core functionality
- Loose IP-based design methodologies
- A clear cultural gap between design and test teams, including "over-the-wall" communication breakdowns

These and many more reasons are today resulting in typical DFT failures, manifested in a variety of forms:

- Lack of strict protocol compliance and loose interoperability
- Deviation from strict functional behavior for test implementation (e.g., accuracy in scan-based delay-path test setup and extraction)
- Poor testability coverage due to logical errors in the implementation (e.g., inability to access BIST controllers or error status reporting registers)

- Decrease in test efficiency (time, test data set size) due to noncoherent test implementation

Motivation

DFT failures due to loose design practices, however, have been commonplace throughout the history of modern IC design. What has changed recently to accentuate the problem? The answer lies in the inevitabilities of test economics. Cost of Test (COT) in the nanometer era is breaking semiconductor economics:

- COT does not scale. Although silicon fab costs have been steadily decreasing to accommodate industry needs, the capital costs of testing wafers have remained flat [1, 2]
- Large Automated Test Equipment (ATE) system cost is driving capital COT, due to complexity of modern SoCs
- Exploding test time and test vector sets combined with low yield are putting immense pressure on COT

In order to deal with the inevitabilities of COT, the industry is beginning to turn to massive DFT implementations:

- Enable low-cost tester deployment by partitioning test resources on-chip
- De-compress test data and compress response data
- Design scalability into test schemes
- Increase controllability/observability for silicon debug
- Implement on-chip instrumentation

These trends are leading to highly complex and sophisticated DFT structures. However, associated methodologies and design practices have not yet caught on to this pressure:

- Although the industry is transitioning to IP-based design to tackle complexity, test is still very flat
- The developing ecosystem of IP vendors and integrators is leading to more heterogeneous and unpredictable test infrastructures
- DFT insertion at different levels of abstraction (RTL, gate, physical) is increasing unpredictability and making it difficult to define QoR requirements

- Test infrastructures are inherently heterogeneous. IP-based design places a new requirement to build coherent system-level test schemes from incoherent components

DFT has hence become too important to treat as a secondary design function and too complex to tackle with traditional approaches. Instead, design teams need to take special care to ensure the behavioral functionality, strict compliance, and efficient operation of their test infrastructures. As silicon test transitions from a design afterthought to a critical manufacturability requirement, companies need to rediscover "Design" in DFT. We start with *verification*.

A Unified DFT Verification Methodology

In trying to design a complete DFT verification environment [3] and associated methodology, one needs to define the key objectives this approach is trying to achieve:

- A well-defined entry points into the design process that can be used as the foundation for expressing test intent and expected end QoR
- Mechanisms for verifying classes of DFT components which will handle the stimuli generation and checking aspects of testing at different levels of abstraction
- Flows for deploying and executing verification as well as measuring progress
- Tracking and analyzing results
- High levels of automation and reuse
- Integration of Test Information Models (TIMs), such as Boundary Scan Description Language or Core Test Language (CTL) files in the verification flow
- Methods for exchanging information with postsilicon applications such as silicon debug and manufacturing test

We have hence designed a robust, unified, DFT-VM. Keeping the stated objectives in mind, we now proceed to describe the methodology based upon three distinct foundations: Planning, executing, and automating.

Planning

The foundation for systematic DFT verification is a well-defined set of goals, supported by a methodology developed to provide integration-oriented test methods into chip-level DFT, enabling compatibility across different embedded cores and incorporating high levels of reuse.

But how can one proactively plan for virtually arbitrary DFT implementations that can be produced by IP-based design, particularly when different vendors follow completely different approaches to DFT? Obviously test plans need to be very modular and reusable, allowing for hierarchical structures to be easily constructed to describe the test infrastructure at hand. Furthermore, test plans need to be polymorphic, very much in the way that object-oriented methodologies define classes of objects, making it possible to use them in a variety of different forms and shapes by specifying simple parameters.

In our solution, we specify a *plan case database*, a repository of plan templates, or *cases*. Such cases contain policies for verifying DFT components such as a JTAG TAP controller [4], without making any assumptions for the nonstandard or implementation-specific aspects of the components. DFT planning cases have the following characteristics:

- They provide blueprints for verifying classes of DFT components
- They specify QoR metrics that verifiers can use to track progress against the plan
- They allow different views into the verification plan data to be specified, allowing for better analysis of results

Planning cases can be used to instrument the verification of both rudimentary DFT components and highly complex structures. This is achieved by *dynamic planning*, the process of hierarchically piecing together a high-level verification plan from lower level plans (see Figure 17.1). This modularity enables the quick and repeatable composition of detailed verification plans for arbitrary DFT infrastructures at the block, core, or system levels. Users can spend time experimenting with these high-level plans for optimum results, setting the blueprint for a well-designed test infrastructure before a single design decision has been made. The ability to reuse plans at different levels of integration and abstraction is a huge benefit to the predictability and verifiability of the project.

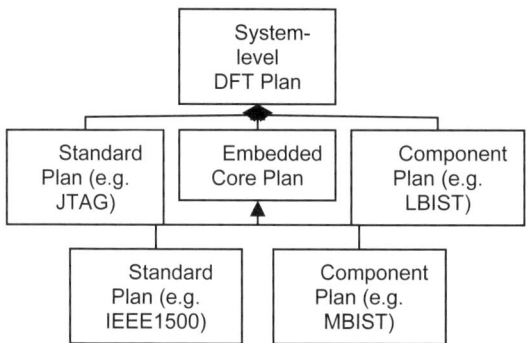

Figure 17.1 Hierarchical DFT Verification Planning

Having compiled a plan case database, we now have the necessary building blocks for expressing high-level, complex, and, most importantly, highly configurable verification plans, maximizing reuse, and leveraging on existing experience. Building a dynamic chip-level DFT verification plan is now broken down to instantiating and configuring multiple DFT verification case objects.

Executing

Once the critical task of planning has been properly addressed, the verification environment needs a scalable way of executing verification on the test infrastructure. The most complete and reusable way to achieve this is by deploying *Verification IP* (VIP).

The concept of verification IP is fairly new in the design community. Conceptually, VIP provides a way of separating generic concepts of design verification from application-specific ones. When this separation is well designed, the direct benefits are enhanced reuse and leverage on existing experience. In the context of DFT, generic concepts can include generating pseudorandom vectors and driving them into a scan chain. Application-specific concepts, for instance, could include using this scan chain to configure a Memory Built-In Self-Test (M-BIST) controller [5].

Essentially, VIP is *mechanism*. It provides the means and capabilities to perform operations and observe DUT behavior; however, it does not include policy. Policy, in this context, is defined as the systematic flow of verifying a complex design, starting with a detailed set of goals, adding a plan of action, and targeting a certain QoR. Hence, starting with a good policy, we can reach our goals by deploying VIP as our mechanism.

We hence define a *VIP Class Database*. This database includes VIP classes which map to types of DFT components such as Test Access Mechanisms (TAMs), scan chain elements, BIST controllers, instruments, etc. Each VIP class includes the tools needed by the verification environment to effectively exercise its corresponding type of DFT logic [6]:

- Constrained random stimuli generators
- Automated, dynamic, checkers, and assertions
- Total coverage collectors

As with plan cases, VIP classes do not include application-specific or implementation-dependent aspects of the DFT component types they target. Rather, they are rudimentary verification environments which are highly reconfigurable and reusable, making it easy to put together complex environments in relatively small time and with reduced effort. Furthermore, the VIP Class Database becomes an experience repository for DFT, where periodic updates ensure uniform design policies and improved interoperability.

Finally, such a repository also helps improve resource utilization and project management. Expert verification engineers can maintain and extend the repository with upgraded capabilities and new functionality while logic designers, usually not entirely familiar with the internal workings of the VIP itself, can simply use the platforms based on its capabilities. Conversely, using this methodology, logic designers can ensure that new features or design changes added directly into VIP classes are made available instantly by regenerating the environment. This enhanced automation of the DFT-VM is discussed next.

Automating

TIMs are schemas used to convey information about the test infrastructure of an IC or embedded core. They typically convey three types of information:

- *Intent*, such as protocol, test modes, etc.
- *Architecture*, such as scan chains, signals, and other design information
- *Data*, i.e., complete test programs that the infrastructure can execute

TIMs hence also serve the purpose of delivering test vectors generated using EDA tools to semiconductor testers (ATEs). The IEEE 1450-1999 Standard Test Interface Language (STIL) [7] is quickly becoming the de-facto standard. Recent extensions to TIM standards (IEEE 1450.1-2005 [8], IEEE P1450.6 [9]) support additional structures in test models to fully describe the DFT architecture itself, thus enhancing the use of such models in semiconductor design environments. These extensions are targeted at enhanced DFT and DFM applications, where ATEs can also be used for analyzing failure data and providing feedback to EDA tools.

In our DFT-VM, TIMs play a significant role. First, TIMs need to be considered a part of the test infrastructure itself. In fact, the recently published IEEE 1500-2005 Standard for Embedded Core Test (SECT) [10] defines a TIM as the only mandatory test infrastructure element for claiming that an embedded core is compliant to the standard. Based on the test intent described in the TIM, designers can provide the necessary functionality while maintaining flexibility in the actual hardware implementation. Hence, TIMs *need to be verified* alongside the DFT components that implement them.

Secondly, TIMs include all the necessary topology, architecture, and implementation-specific information that must be available to the verification environment. This way, a silicon IP vendor can communicate test intent of a design core to an integrator within specified completeness, interoperability, and confidentiality requirements. This information, in the form of a TIM, can then be used by the integrator for a variety of design functions ranging from implementing certain DFT components to shaping the IC-level test infrastructure (Figure 17.2).

In our approach, we are extending the applicability of TIMs to design verification, claiming that TIMs can provide an automation link between DFT design and verification. The argument is supported by a variety of technical and business conditions:

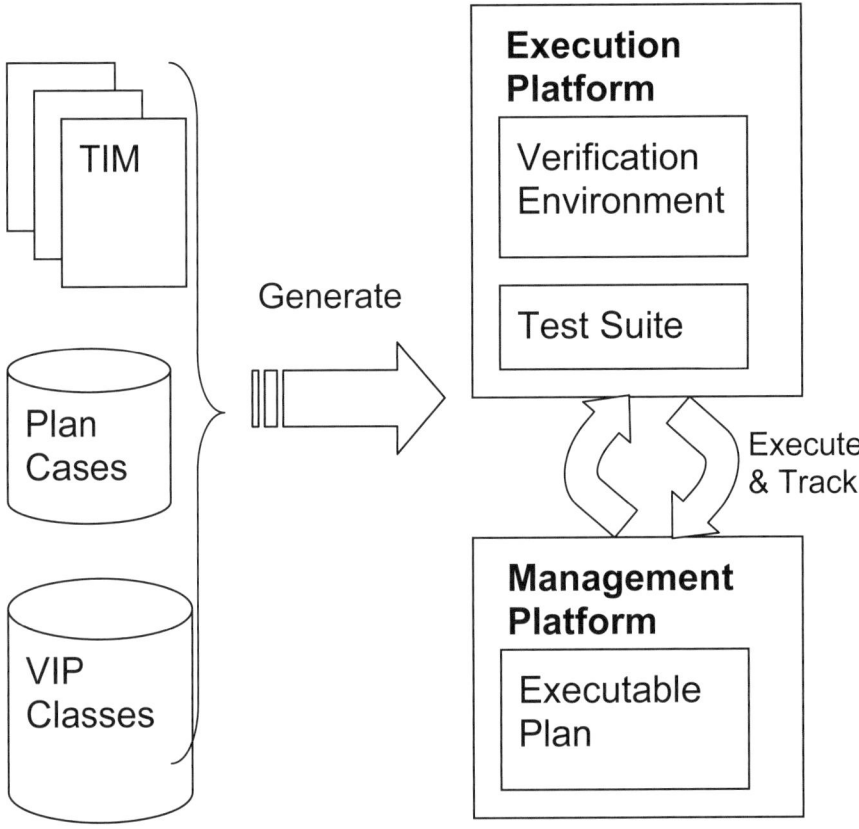

Figure 17.2 DFT-VM Automation Flow

17 Design for Test

- TIMs can encompass test intent as specified by test engineers without committing to design decisions and hence provide the grounds for an early test specification
- TIMs bridge the gap between IP vendors and integrators with respect to DFT support in IP cores and hence can be used to verify deliverables
- TIMs are models that can be used for early test performance exploration
- Verification automation based on TIMs can be used to maintain a link between the post- and presilicon worlds, allowing testbenches to be reused for debugging silicon and optimizing manufacturing test

In order to better understand how TIMs can be used to build DFT-verification environments quickly and effectively, let's consider the following example in the context of an IEEE 1500-2005 compliant embedded core.

Test Case

IEEE 1500-2005 (SECT) defines a scalable architecture for independent, modular test development, and test application for embedded design blocks. It also enables test of the external logic surrounding these cores. Modular testing is typically a requirement for embedded nonlogic blocks, such as memories, and for embedded, predesigned, nonmergeable IP cores. In addition, the IEEE 1500 architecture can also be used to partition large design blocks into smaller blocks of more manageable size and to facilitate test reuse for blocks that are reused from one SoC design to the next.

A typical TIM is that of an IEEE 1450.6 CTL description of the IEEE 1500 test infrastructure, commonly referred to as a wrapper, found in a SECT compliant embedded core (for more information please refer to the IEEE 1500 standard). Such a model includes, amongst others, information in a parse-able format about:

- *Signals.* The TIM publishes information about signal names and sizes, as well as their default state, so they can be initialized and driven/sampled by an external agent.
- *Scan Chains.* This information refers to scan structures that are part of DFT. That includes scan chain sizes and cells' names, so other information regarding the cells like parallel inputs and/or output connecting signal names can also be inferred.
- *Scan Cells.* TIMs publish all information pertaining to the scan chain cells, since IEEE 1500 cells follow a standard naming convention that fully describes their structure and function.
- *Test modes.* The TIM also includes information about the various test modes that can be reached by loading appropriate instructions. It provides the instruction opcode that triggers this mode, identifies the data register to be used, and provides the macros used to access it.

A TIM parser can parse all this information and infer:

- IEEE 1500 control signals:
  - The instruction set used
  - Opcodes
  - Data registers referenced
- The collection of test data registers:
  - Sizes
  - Signal connections
- The cells contained in those registers:
  - Structure
  - Signal connections
  - Behavior during capture, update, transfer operations

As an example, Figure 17.3 illustrates an example of a CTL description for an IEEE 1500 compliant embedded core wrapper. The description provides information about the wrapper, including the size and cell type of the instruction register (WIR).

17 Design for Test

```
// CTL for IEEE 1500 enabled embedded
   core wrapper
...
ScanChain wir_chain{
  ScanLength 4;
  ScanCells wcell[0..3];
}
...
  ...
                      Internal{
    'wir_fi[0..3]{ DataType Functional
                 TestData;
      IsConnected In {
        StateElement Scan 'wcell[0..3]';
        ..
      }
                              }
  ...
  }
  ...
}
```

Figure 17.3 IEEE 1450.6-CTL DFT Structure Example

Having extracted these structures from the CTL model, one can envision a process by which:

- The corresponding IEEE 1500 VIP class is selected from the VIP database and instantiated (see Figure 17.2)
- The number of cells specified in the WIR structure and signals connecting to the parallel input of those cells are used to configure the VIP (see Figure 17.4)

```
// 'e' language configuration for an
   IEEE 1500 VIP module
extend WIR glbt_sect_ref_model_register {
  keep size == 4;
};
      extend WIR glbt_sect_ref_mode_register_cell
                           {
  keep cfi == append("wir_fi", "[",
                cell_index, "]");
};
```

Figure 17.4 Environment Generation Based on a TIM

Collections of TIMs can be grouped together hierarchically to perform system level DFT verification. This can be done by analyzing the TIMs and deducing the respective topology of each embedded core and its corresponding DFT infrastructure in the SoC. With this information, DFT-VM can be used to dynamically create testbenches and tests optimized for a specific DFT configuration.

Benefits

The described verification methodology serves as a solid foundation for true design for test. By enforcing early verification documentation and planning, it aligns the perspective of different design teams with respect to DFT support and enhances visibility. Automating environment generation, it ensures that logic designers and test engineers have a good auditing system for debugging and regression analysis, while propagation of new features and updates is centralized through the use of plan and VIP databases. Better project management and more efficient resource utilization are also achieved by providing clear interfaces for logic designers and verification engineers.

TIM coverification introduces strong semantics into the description and integration of test infrastructures. DFT designed by separate teams or IP vendors can be merged into the IC-level reliably, while maintaining a link with manufacturing test deliveables. Architectural changes to DFT can quickly propagate to the design environment through fast regeneration and automatic plan updates. Vendor qualification for DFT becomes possible by enforcing TIM deliverables and being able to quickly and reliably validate vendor claims for testability and interoperability. Finally, advanced DFM applications can also be supported through early collaboration with the fabrication and tester providers.

Enhanced design engineering, automation, and reuse lead to increased predictability, better productivity, and higher overall quality.

Future Work

Having successfully applied DFT-VM to functional verification, we are envisioning several other areas where the methodology can scale:

- *Overall validation of test deliverables*
  The methodology described herein successfully sets the foundation for verifying the functionality, compliance, and completeness of DFT netlists. The next step would be to extend the methodology to include coverification of test sets with DFT netlists, allowing engineers to incrementally verify and optimize all deliverables of a comprehensive, systematic, test flow. Validation of test intent would also encompass performance aspects of semiconductor test, such as power profiling, ATE constraint analysis and related cost optimization.
- *Postsilicon test and validation*
  Having created a completely DFT aware verification environment with QoR measurements and associated test sets, postsilicon validation of DFT becomes a natural step in the methodology. Validating DFT in silicon in a systematic and predictable manner can help save test time and improve reliability in manufacturing test.
- *DFM applications*
  It is also intended to extend the methodology to some interesting DFM applications such as importing TIM-based test results from silicon test back to verification to gain better understanding of the functionality perspective of common failures and to facilitate analysis and debug.

Conclusions

We have identified the need to systematically verify DFT as part of the total system verification process, in order to increase the quality of the design and by virtue, the end product. The need becomes more apparent in the context of IP-based design of SoCs, where multiple embedded cores from different providers introduce heterogeneity and variation of DFT quality.

We have hence proposed a unified methodology for DFT verification, using TIMs to dynamically identify, instantiate, and configure executable verification plans and environments. Hence we provide a fast and reliable way to building automated testbenches capable of verifying DFT designs from simple components to complete test infrastructures. Our approach enables true design for test based on measurable QoR, and enhances productivity, reliability, and reusability.

Finally we have demonstrated how our methodology results in a verification infrastructure that can be reused during silicon debug and test vector design for several advanced applications, forming the basis for future work.

References

1. International Technology Roadmap for Semiconductors, 2003 Edition, "Test & Test Equipment"
2. International Technology Roadmap for Semiconductors, 2004 Update, "Test & Test Equipment"
3. K. Melocco, H. Arora, P. Setlak, G. Kunselman, and S. Mardhani, "A Comprehensive Approach to Assessing and Analyzing 1149.1 Test Logic," in the proceedings of International Test Conference, Charlotte, NC, USA, September 30–October 2, 2003, pp. 358–367
4. IEEE Computer Society, "IEEE Standard Test Access Port and Boundary-Scan Architecture – IEEE Std. 1149.1-2001," New York: IEEE, 2001
5. D. Appello, F. Corno, M. Giovinetto, M. Rebaudengo, and M. Sonza Reorda, "A P1500 compliant BIST-based approach to embedded RAM Diagnosis," in the proceedings of 10th Asian Test Symposium, Kyoto, Japan, November 19–21, 2001, pp. 97–102
6. I. Diamantidis, T. Oikonomou, and S. Diamantidis. "Towards an IEEE P1500 Verification Infrastructure: A Comprehensive Approach," presented at the 3rd IEEE International Workshop on Infrastructure IP (IIP), Santa Clara, CA, USA, May 4–5, 2005
7. Test Technology Standards Committee of the IEEE Computer Society, "IEEE Standard Test Interface Language (STIL) for Digital Test Vector Data – IEEE Std. 1450-1999," New York: IEEE 1999

8. P1450.1 Working Group of the Test Technology Standards Committee, "Draft Standard for Standard Test Interface Language (STIL) for Digital Test Vector Data – Extensions to STIL for Semiconductor Design Environments – P1450.1," New York: IEEE 2005
9. CTL Working Group of the Test Technology Standards Committee, "Draft Standard for Standard Test Interface Language (STIL) for Digital Test Vector Data – Core Test Language (CTL) – P1450.6/D1.6," New York: IEEE 2005
10. IEEE Computer Society, "IEEE Standard Testability Method for Embedded Core-based Integrated Circuits – IEEE Std. 1500-2005," New York: IEEE 2005

Part IV
Case Studies and Commentaries

Metric-Driven Design Verification: Why Is My Customer a Better Verification Engineer Than Me?

Alfonso Íñiguez, Freescale Semiconductor Inc.

Alfonso Íñiguez is a principal staff verification engineer with the Security Technology Center at Freescale Semiconductor, where he is the verification lead responsible for developing, improving, and applying functional verification tools and methodologies. His work includes cryptographic hardware accelerator design. He holds a B.S. in Computer Engineering from the Universidad Autónoma de Guadalajara, México, and an M.S. in Electrical Engineering from the University of Arizona. Due to his dyslexia, Alfonso did not learn how to read a full sentence until he was 18 years old and he still does not know how to subtract. Alfonso commutes to work by bicycle averaging 90 miles per week. He is a large format photographer, bongo player, and salsa dancer. He presently lives in Mesa, Arizona with his wife, three children, and many chickens.

Abstract

Why is it that after months of directed and random testing you were not able to find a bug that your customer found within two days of receiving samples? Is there anything wrong with your directed and random testing? Should you blame it on faulty assertions? Could it be that you did not run your simulation long enough? Could the bug have been discovered by using better coverage criteria? The intention of this paper is to answer all those questions by analyzing past mistakes and proposing an effective way of writing a thorough metric-driven verification plan.

This document is a compendium of experiences, containing verification pitfalls and prevention strategies, which the author has witnessed throughout his 15-year career in the fields of product evaluation, applications, design, and verification engineering.

Traditionally verification plans are written with the design specification in mind. There is nothing wrong with this approach. In fact, it is an essential requirement, but evidently, this common practice is not sufficient. There is an important component called software validation, which is traditionally left to the software team to complete. Two scenarios can be described for software validation. The software team completes this step by using a model of the DUV or it uses the traditional silicon evaluation board approach. The first approach is highly recommended, but since the team is working with a software model of the DUV, it may overlook signal contention, race conditions, and a myriad of other timing problems. If the software team chooses the silicon evaluation option, then the company is relying on a very expensive debugging methodology, with the risk of an exorbitant shift in the delivery schedule and lost of credibility in the design team. To avoid those problems, a software validation approach should be included in the functional verification process.

Introduction

Most verification publications start with the following suggestion: "first write a verification plan" which by the way is an excellent suggestion. However, when writing a verification plan you need to consider past mistakes. The collection of verification pitfalls described in this paper should serve as a preamble to writing an effective verification plan.

Why was my customer able to find a bug that I overlooked? I can assure you that I can come up with a large number of convincing explanations, but no matter how creative the explanations are, they always fall into one of the following two categories:

    (a) The customer and I had different definitions of the intended functionality.

(b) The customer was the one who set the delivery schedule, not me.

Although "a" and "b" are true, please do not use those reasons as excuses when things go wrong. I have listed them here as a starting point in finding a solution to the problem.

Section 1: The Elusive Intended Functionality

In order to understand the intended functionality of a given DUV, it is necessary to define who the customer is. As we will see in the next section, the definition of customer can be very extensive.

Defining the Customer

To a functional verification engineer, usually the first thing that comes to mind when we use the term "customer" is an external company building a PDA, cell phone, network card, computer, or any other finished product. The term customer should not be limited to that external company, but rather extended to anyone who is capable of spoiling your weekend. Your customer pipeline begins with the designer of the DUV, and extends all the way up to the end user who has found new ways of using the product and knows more about the capabilities of the design then the designer himself. A more specific definition of customer includes the following people: the IP Designer, the IP integrator, the software developer, the marketer, and the end user. Now that we know who the customer is, we can proceed to define the intended functionality per customer.

The IP Designer as Customer

You, the verification engineer, are a service provider, and like in any other profession, the best service providers are the ones who get their customers involved in the process. For example, a good surgeon can increase her success rate by persuading her client to diet and exercise before the surgery, which is something that the surgeon cannot do for her client. Similarly, the verification engineer needs to get her client, in this case the designer, involved in the verification process. Such involvement is the remedy to prevent the following two pitfalls:

(a) Unreviewed verification plan.
(b) Unwritten white-box assertions.

Verification Pitfall #1: Unreviewed Verification Plan

Nowadays, a typical verification engineer is on a race against the clock, which could lead to corner cutting the review of the verification plan. If you have already spent a week writing a verification plan, please spend an extra day reviewing the plan with the design team. Undermining the importance of the review could result in costly silicon respins. In my personal experience, failing to review the plan's random constraints triggered the following consequence: A random test case verified a DMA block by scattering data throughout memory using hundreds of links, but the random constrain section of the test, failed to include simple scenarios that used only one or two links, which is where the bug was hidden. In this case, the chip-level integrator, who found the bug, demonstrated to be a better verification engineer than me.

Verification Pitfall #2: Unwritten White-Box Assertions

Not all assertions are meant to be written by the verification engineer, such is the case of the white-box assertions embedded in the RTL, which should be owned by the designer. At least every state machine, FIFO, data pipeline, and instruction pipeline should have assertion checkers. I have recently encountered a condition in which the FIFO, once full, delayed the assertion of its full signal by one cycle. Currently, this bug has not caused a problem in the field because the surrounding logic is incapable of reenacting the failing scenario, but if the FIFO is used as IP on a different design, then the bug could appear. This could have been prevented by writing a simple white-box assertion.

The IP Integrator as Customer

Two types of potential problems come to mind when I put myself into the IP integrator's shoes, the possible verification pitfalls are:

(a) Extending the definition of false-bugs.
(b) Using adjacent blocks as checkers.

Why Is My Customer a Better Verification Engineer Than Me?

Verification Pitfall #3: Extending the Definition of False-Bugs
False-bugs are the infinite number of possibilities that the DUV will never see because of the nature of the interface. Here is an example of a false-bug: The DUV generates unpredictable data when a glitch is injected to the address line. Is this a real bug? I do not think this is a bug, because the interface specification does not specify this kind of noise scenario. Think of the interface specification as an insurance policy that protects the verification engineer from an unlimited number of false-bugs, see Figure 1.

A common verification pitfall is to overextend the definition of the false-bugs by using adjacent IP blocks as a false-bug protector, see Figure 2. The adjacent IP might provide a protection on a given platform configuration, but leave the DUV unprotected once it is integrated into a different platform configuration.

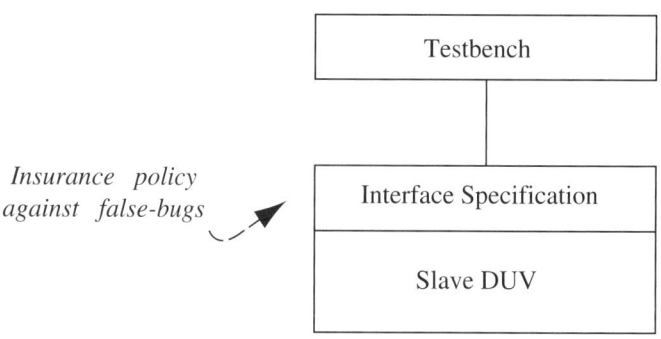

Figure 1 The Interface Specification Protects the Design from an Unlimited Number of Absurd Signal Combinations

Verification Pitfall #4: Using Adjacent Blocks as Checkers
A BFM is capable of generating cycle-accurate signals as described by a given bus protocol. The most elementary purpose of a BFM is to verify a slave DUV by emulating a host processor connected to its interface.

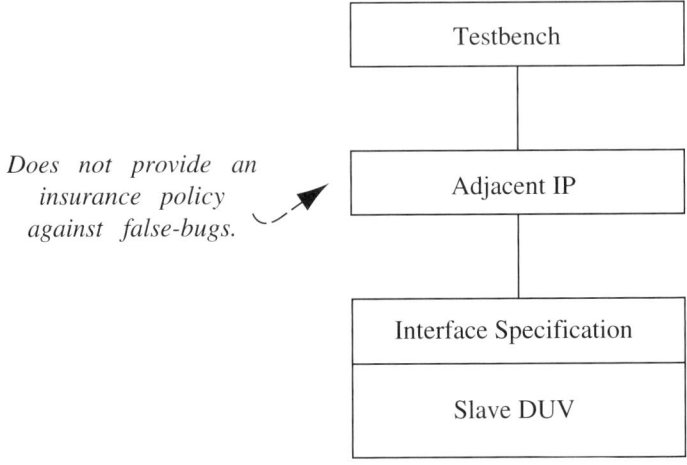

Figure 2 An Adjacent Block Provides a False Sense of Security, Since it Does Not Protect Against False-Bugs

Figure 3 shows the simplified data flow between a BFM and a slave DUV:

(1) The BFM writes data into the DUV.
(2) The BFM reads the result from the DUV.

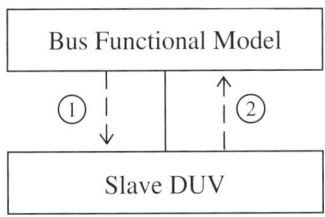

Figure 3 Data Flow Between a BFM and a Slave DUV

There is nothing wrong with this type of verification strategy. However, when it comes to verifying a master DUV, the BFM faces a fundamental limitation. If we connect a BFM directly into a master DUV, their respective driving signals would collide. This collision can be avoided by adding bus arbitration logic between the BFM and the DUV, but that would only solve part of the problem. In order to verify a master DUV effectively, we need to construct a platform. Figure 4 shows a simple platform and the data flow between the BFM and master DUV:

(1) The BFM writes the object code into the RAM.
(2) The BFM gives an execution command to the master DUV.
(3) The master DUV begins executing the RAM's object code.
(4) The master DUV writes the result into the RAM.
(5) The master DUV sends a "done interrupt" signal to the BFM.
(6) The BFM reads the result from the RAM.

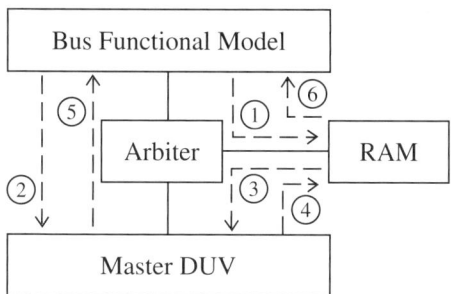

Figure 4 Data Flow Between a BFM and a Master DUV

The verification pitfall in this strategy is in using adjacent blocks as checkers. In the example above, the arbiter block, regardless of its IP quality, should not be used as the only criteria to test the DUV's interface, we must introduce an independent bus monitor to verify the bus protocol, see Figure 5.

In my personal experience, failing to use a bus monitor led to the following undetected bug: The master DUV equipped with an AMBA AHB interface failed to "walk" a burst when crossing a 2K boundary (to "walk" a burst is to transition from sequential to nonsequential transaction, as described in the AMBA specification).

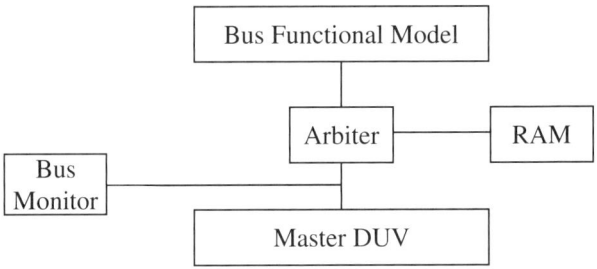

Figure 5 The Bus Monitor is Indispensable for Bus Protocol Verification

The Software Developer as Customer

Take a software validation approach by putting yourself into the software developer's shoes; this will prevent you from falling into the following two verification pitfalls:

(a) Using the design specification as the only model to generate stimuli.
(b) Restricting the stimuli to valid data.

Verification Pitfall #5: Using the Design Specification as the Only Model to Generate Stimuli

Even using a software driver's recipe as a guideline when building a stimulus generator is insufficient because in the life time of a given design, the software driver is likely to be upgraded multiple times to increase functionality and improve performance. Each time the software driver is revised, the order in which transactions are executed might change. For this reason, when building a stimulus generator, the verification engineer must build a dynamic software driver that considers variable execution sequences.

Verification Pitfall #6: Restricting the Stimuli to Valid Data

When writing tests or stimulus generators, we tend to concentrate on predictable test case scenarios. For example, assume that a design specification defines a 2-bit register with the following valid modes: 2'b00, 2'b01, 2b'10, and then it defines the value 2'b11 as reserved.

How would you normally verify that register? You would probably create test case scenarios for all the valid modes (2'b00, 2'b01, and 2b'10) and if it passes you are done. I have used this strategy in the past until one day the untested "reserved"' value, set by the software driver, deadlocked the design. The moral of this story is that you must always verify all reserve bits and invalid modes on all Registers.

The Marketer as Customer
By using the marketer perspective, we can uncover an important verification pitfall: Not testing for performance.

Verification Pitfall #7: Not Testing for Performance
The performance of a given product is traditionally measured during silicon evaluation. There is a problem with this methodology. Disappointing silicon evaluation results could spoil the launch of a new product, sometimes ending in product cancellation.

From the point of view of functional verification, determining performance by measuring data throughput is feasible. Allow me to congratulate you if you are already including data throughput tests in your verification plan. Technically speaking, not meeting expected performance is not a functional bug. However, knowing this information ahead of time can lead to a timely design change or conclude that the expected performance is unrealistic and unobtainable.

The End User as Customer
By using the end user perspective, the following verification pitfalls come to mind:

> (a) Restricting the verification plan by the limitations of the testbench.
> (b) Overlooking gate-level simulation.
> (c) Halting the simulation once full functional coverage is reached.

Verification Pitfall #8: Restricting the Verification Plan by the Limitations of the Testbench

The best verification plans are writing by engineers who are not familiarized with the testbench. As a verification engineer, you must be familiarized with the limitations and capabilities of the testbench, but do not use that information when defining the test cases. Put yourself into the end users' shoes. Think about possible unintended use cases, experiment with error conditions, and see if you can recuperate gracefully without having to use the reset signal. Once the test cases are defined in the verification plan, make the proper modifications to the testbench to make it capable of replicating those scenarios.

Verification Pitfall #9: Overlooking Gate Simulation

LEC (Logic Equivalence Checking) ensures that the functionality of the gate-level netlist matches the RTL. STA (Static Timing Analysis) quickly examines clocking schemes and identifies timing problems up front. It can analyze multiple conditions in a single run, dramatically reducing gate-level verification time.

Does it mean that we can replace gate-level simulation by running LEC and STA instead? In my personal experience, the answer is "no." STA is especially prone to error, since it requires human intervention to classify true and false paths. On a simple design, all offending data paths can be accurately classified, but on a complex design, the number of paths can be overwhelmingly high and prone to bad judgment. This is not because the designer is incompetent, but because human attention span is intermittent by nature.

With the arrival of large designs, rerunning the RTL tests on a gate-level simulation has become prohibitive, but that should not prevent us from running a selective subset of the RTL tests on a gate-level simulation.

Verification Pitfall #10: Halting the Simulation Once Full Functional Coverage is Reached

For practical purposes, it is necessary to define a full functional coverage criterion. Although, on a complex design, full functional coverage is practically unobtainable. Regardless of how much computing power and how much available time you have, you can be certain that your end users will execute millions more tests than you ever will. For this reason, you should not stop simulating once your coverage goal has been reached, in fact, even after tape-out you should keep running your regression for a prudent time.

Once full functional coverage has been reached, readjust the random constraints and rerun the regression. Keep in mind that running diverse and meaningful tests is more important than just running the same test with different values.

Section 2: The Ever-Shrinking Schedule

Nowadays, almost every verification article or verification tool sales pitch starts with the following line: "functional verification takes 70% of the chip design cycle." Is this guideline accurate? Unfortunately, the 70% rule is only taken in to account at the beginning of the project, and soon after everybody ignores its existence.

For the sake of argument, assume that the 70% rule is actually followed by every project member. In that case, a hypothetical dialogue between the project lead and the verification engineer could go as follows:

Project lead: *How much time do you need to verify this design?*

Verification engineer: *It depends how long would it take to design it?*

Project lead: *Well, I've just spoken to the designer; she said it would take three months.*

Verification engineer: *(The engineer takes a few seconds to apply the 70% rule and responds) if the design takes three months to design it, then I will need seven months to verify it.*

Project lead: *Very well, let me give our proposed schedule to our customer.*

Verification engineer: *Wait; there is something else that you need to know. Once the schedule is finalized, if the customer makes changes the design specification, then I will need to adjust our delivery schedule using the 70% rule.*

Project lead: *Very well, I will make sure that our customer is aware of the 70% rule.*

The only thing wrong with this conversation is that it can only happen in verification haven, it will never occur in the real world. In my view, the 70% rule is a nice guideline that makes good academic schedules, but the real schedule is more likely to be determined by the competitive nature of the semiconductor industry.

A more realistic dialogue between the project lead and the verification engineer would go like this:

Project lead: *We have been presented with the opportunity to deliver this new design to one of our most important customers.*

Verification engineer: *Okay.*

Project lead: *The challenge is that the customer needs to have samples by the end of the year.*

Verification engineer: *I assume you are referring to the end of next year, right?*

Project lead: *No, they need them by the end of this year; we have six months to deliver.*

Verification engineer: *That would be impossible.*

Project lead: *The customer is aware of the aggressive schedule; to make it easier for us they have agreed on removing functionality X and Y from the design.*

Verification engineer: Well, even if we remove functionality X and Y, it is still aggressive, we must move the delivery date to eight months from today instead of six.

Project lead: Okay, that sounds reasonable.

One day later:
Project lead: The customer agreed on our proposed schedule. We will deliver samples in eight months.

Verification engineer: Good.

A month later:
Project lead: The customer said that they absolutely need to have functionality X. They initially said that they could live without it, but now they are asking for it, the good news is that the customer promised to buy twice as many parts as initially promised.

Verification engineer: Well, I am going to need more time to verify functionality X.

Project lead: That is okay, we have already negotiated more time for you, and we have added an extra 15 days to the delivery date.

Engineer: But I am going to need a lot more time to write test cases and assertions.

Project lead: Well, let us just concentrate on the test cases and leave the assertions for the next project.

Although, this dialogue is fictitious, it is not too far away from the reality of the semiconductor industry. At the end, the pressure to survive in this competitive industry determines the schedule and forces the verification engineer to cut corners. The verification engineer should speak up and preserve the quality of his or her work by writing a metric-driven verification plan. The verification plan is a statement of work; you are going to be in a better position to negotiate a more reasonable schedule if you have this document. If you currently feel overworked and stressed in your present project, maybe it is because you have not taken the time to write a metric-driven verification plan.

The Universal Remedy to an Oppressing Schedule: The Metric-Driven Verification plan

I know very few engineers who actually enjoy writing document-tation. I am the first one to admit that it is not fun, but if you are not willing to overcome this problem then your customer will always be a better verification engineer than you. The structure of the verification plan is a matter of personal style and sometimes company's style. However, it does not matter what the style it is as long as it includes the following sections:

- (a) Testbench architecture
- (b) Platform architecture
- (c) Directed test cases
- (d) Random test cases
- (e) Failure test cases
- (f) Coverage, FV, STA, BIST, SCAN pre-, and postlayout simulation

Testbench Architecture

The topic of the testbench architectures is broad enough to fill up an entire book. Besides, it is not the intention of this document to cover it. Assuming that you have already chosen an architecture for the testbench, make a block diagram and include it in this section with a brief explanation of each block, but not include implementation details. The purpose of this section is to familiarize your reader with the testbench and your terminology.

Platform Architecture

If you are verifying a block with master capabilities, then you need to build a platform, as described in Section 1.3.2. Such platform needs to be reconfigurable. Figure 6 shows a brief example of a configurable platform, which includes a port swapping construct using the *'ifdef* Verilog directive.

```
'ifdef DUV_ON_PORT_1
        // BFM on port 0
        .m0_haddr    (haddr),
        .m0_hwdata   (hwdata),
        // DUV on port 1
        .m1_haddr    (duv_haddr),
        .m1_hwdata   (duv_hwdata),
'else
        // BFM on port 1
        .m1_haddr    (haddr),
        .m1_hwdata   (hwdata),
        // DUV on port 0
        .m0_haddr    (duv_haddr,
        .m0_hwdata   (duv_hwdata),
'endif
```

Figure 6 The DUV and the BFM can Swap Ports by Using the 'Ifdef Construct

Directed Test Cases

These are sanity checkers, i.e., reading the initial value of all registers after reset, or conformance testing using predetermined input and output data. The directed test cases should verify all the basic features of the DUV. All these tests are to be executed and debugged before performing random testing. A subset of the directed tests, if not all, can be ported to the chip level.

Random Test Cases

The entire software validation tests should be included here. Now is the time to beat the design from every angle, the designer's input is crucial, get them involved when reviewing the plan. The random simulation should continue even after tape-out.

Failure Test Cases

Examine the ten verification pitfalls and make sure that each potential pitfall is covered. This is your opportunity to get creative, and challenge yourself to break the design. Verify if the design recuperates gracefully when given erroneous conditions such as interruption of burst transfers, invalid modes, uninitialized external memory, out of sequence register configuration, and out of sequence operations. Feel free to use the "force" statement, but use it with caution. Always avoid creating invalid false-bugs.

Coverage, FV, STA, BIST, SCAN, Prelayout, and Postlayout Simulation

Each one of the disciplines listed here is too broad to be described in this document. Nevertheless, you should specify a strategy for each one in your plan. If BIST and SCAN are outside your jurisdiction, then, take the time to write it in the plan and specify who will be responsible for performing those tasks.

Section 3: Writing a Metric-Driven Verification Plan

This is your opportunity to put an end to the oppressing schedule and increase the quality of your work. When writing the verification plan take time to include metric-driven tasks. Once all the preliminary information has been defined, i.e., testbench architecture, platform architecture, directed test cases, random test cases, failure test cases, and coverage criteria, the next step is to estimate the time necessary to complete each of these tasks.

Estimating the Time Required to Complete Metric-Driven Tasks

The example shown in Table 1 assumes a relatively simple design that is being verified by a single engineer. The tasks can be easily expanded to incorporate complex designs with numerous team members.

Notice the addition of a contingency plan, which is to be used in case of a design change. Be reasonable, if the design change is simple, such as a register remapping or bit redefinition, it would take a longer time to update the metric-driven task table than it would take to make the change to the test case. Do not burden yourself or the team with bureaucratic documentation. Only significant design changes merit a change in the contingency section.

Guidelines for constructing the metric-driven task table:

(a) The testbench development task includes finding or developing infrastructure IP such as: BFM, Monitors, and preverified adjacent blocks.
(b) Keep it simple; use a "workweek" as the smallest unit of measurement. This practice makes schedules easier to develop, modify, and track.
(c) Assertions are included in the test case development tasks. You may create a separate task depending on the complexity of the design.
(d) The design specification and verification plan documents shall have the same version number.

Determining the Delivery Schedule

Once the metric-driven tasks have been defined, make a delivery schedule similar to Table 2, this will help track your progress. This example assumes a single engineer completing each task in a sequential fashion. Make sure that project lead is aware of the requirements to initiate each task.

Table 1 Initial Metric-Driven Task Table

Tasks	Time	Comments
Verification plan 1.0 development	1 week	*Applicable only to design spec 1.0*
Testbench development	2 weeks	*Applicable only to design spec 1.0*
Directed test case development	2 weeks	*Applicable only to design spec 1.0*
Random test case development	4 weeks	*Applicable only to design spec 1.0*
Failure test case development	2 weeks	*Applicable only to design spec 1.0*
Debug RTL 1.0 & test cases	4 weeks	*Applicable only to design spec 1.0*
Regression testing & coverage analysis	4 weeks	*Applicable only to design spec 1.0*
Contingency Plan		
Update verification plan to verify design change	x	*To be added to schedule in case of design change*
Update directed test cases to verify design change	x	*To be added to schedule in case of design change*
Update random test cases to verify design change	x	*To be added to schedule in case of design change*
Update failure test cases to verify design change	x	*To be added to schedule in case of design change*
Write new test cases to verify design change	x	*To be added to schedule in case of design change*
Debug RTL $x.x$ & updated test cases	x	*To be added to schedule in case of design change*
Regression testing & coverage analysis	x	*To be added to schedule in case of design change*

Table 2 Determining and Tracking the Delivery Schedule

Requirements to Initiate Task	Tasks	week 1	week 2	week 3	week 4	week 5	week 6	week 7	week 8	week 9	week 10	week 11	week 12	week 13	week 14	week 15	week 16	week 17	week 18
Design spec 1.0 complete and reviewed.	Verification plan 1.0 Development	■	■																
	Testbench development			■															
	Directed test case development				■	■													
	Random test case development						■	■	■	■									
	Failure test case development										■	■							
RTL 1.0 completed	Debug RTL 1.0 & Test cases												■	■	■				
	Regression testing & coverage analysis															■	■	■	■

Section 4: Implementing the Metric-Driven Verification Plan

If a design change is large enough to put the delivery schedule at risk, update the contingency section of the metric-driven table, see example on Table 3.

Triggering the Contingency Plan

Remember that the minimum measurement unit is one workweek, this is necessary to keep your metrics manageable.

Table 3 Making Use of the Contingency Plan

Tasks	Time	Comments
Update verification plan to verify design change	1	*Additional time to verify design spec 2.0*
Update directed test cases to verify design change	0	*Additional time to verify design spec 2.0*
Update random test cases to verify design change	2	*Additional time to verify design spec 2.0*
Update failure test cases to verify design change	1	*Additional time to verify design spec 2.0*
Write new test cases to verify design change	1	*Additional time to verify design spec 2.0*
Debug RTL 2.0 & updated Test cases	2	*Additional time to verify design spec 2.0*
Regression testing & coverage analysis	2	*Additional time to verify design spec 2.0*

Adjusting the Delivery Schedule

Adjust the schedule according to the contingency plan, see Table 4. In Table 4, the cumulative time required to verify version 2.0 have shifted the delivery schedule by nine weeks.

Table 4 Shifting the Delivery Schedule According to the Contingency Plan

Requirements to Initiate Task	Tasks	Week 1	Week 2	Week 3	Week 4	Week 5	Week 6	Week 7	Week 8	Week 9
Design spec 2.0 complete and reviewed	Update verification plan to verify design Spec 2.0	■								
	Update directed test cases to verify design change	■								
	Update random test cases to verify design change		■	■						
	Update failure test cases to verify design change				■					
	Write new test cases to verify design change					■				
RTL 2.0 completed	Debug RTL 2.0 & updated Test cases						■	■		
	Regression testing & coverage analysis								■	■

Table 5 Collecting Metrics for Future Reference

Tasks	Estimated Time	Actual Time	Comments
Verification plan 1.0 development	1 week	1 week	
Testbench development	2 weeks	4 weeks	Had to write custom monitor due to nonstandard interface usage.
Directed test case development	2 weeks	2 weeks	
Random test case development	4 weeks	6 weeks	Compute farm unavailable due to higher priority project.
Failure test case development	2 weeks	2 weeks	
Debug RTL 1.0 & test cases	4 weeks	4 weeks	
Regression testing & coverage analysis	4 weeks	6 weeks	Had to write additional test to meet functional coverage
Total time to verify design spec 1.0	*19 weeks*	*25 weeks*	
Update verification plan to verify design change	1 week	1 week	
Update directed test cases to verify design change	0 week	0 week	
Update random test cases to verify design change	2 week	3 week	Had to modify testbench infrastructure, need to account for this task next time.
Update failure test cases to verify design change	1 week	1 week	
Write new test cases to verify design change	1 week	2 week	Underestimated complexity.
Debug RTL 2.0 & updated Test cases	2 week	1 week	
Regression testing & coverage analysis	3 week	3 week	
Additional time to verify design spec 2.0	*10 weeks*	*11 weeks*	

Collecting Metrics

The most important part in keeping metrics is recording the actual time it took to complete each task, see column 4 on Table 5. If your estimated time and actual time do not match, then write the problem that provoked the shift in the schedule.

Once you have completed your first project, you will be in a better position to accurately estimate the time for the next one. As a rule of thumb, the task completion time can be made proportional to the number of gates in the design. Always factor in the time benefit of reusing testbench and test case components. Armed with this information, you can easily extrapolate to determine the schedule of more complex designs.

Conclusion

When proposing a schedule, the 70% rule gives us a good guideline, but as new changes are added to the design, the final percentage dedicated to verification devaluates to 60%, 50%, or 40%. Regardless of market pressure, the time dedicated to verification should never decrease to the point in which you are shipping untested logic. It is your responsibility to write a metric-driven verification plan and use it to negotiate a reasonable schedule.

In my view, the most difficult part of the plan is predicting the time required to complete each task. Even if you become proficient at estimating the time required to write a specific test, you still need to factor in an elusive component, which is the debugging time. From experience, you have probably already discovered that some problems can be debugged within minutes, but others might take days to resolve. Keep a log of the time spent writing tests and debugging the DUV, and use this information to estimate schedules of future projects.

As a professional in this field, you should execute the verification plan to its end. If a feature is untestable due to some unforeseeable limitation, document it on the plan. If you are asked to cut corners to speed up the sign-off process, use your plan to inform your management and make them aware of the risks they are taking by compromising the verification process.

Metric-Driven Methodology Speeds the Verification of a Complex Network Processor

Jean-Paul Lambrechts, Cisco Systems

Jean-Paul Lambrechts has over 20 years experience in leading hardware design in the networking and computer areas. His experience covers board-level hardware design, FPGA, and verification. Jean-Paul has now been with Cisco for 9 years where he has been responsible for line cards, packet forwarding engines, and layer 4–7 processor card. Jean-Paul holds a MSEE degree from the Louvain University in Belgium.

In my business unit, we often struggle to keep our hardware in sync with the software schedule. Doing so requires a great deal of resources. So it was no small feat when the verification flow for one of our recent designs, the Programmable IP Services Accelerator (PISA) FPGA in development, was completed ahead of schedule and well before the system software was delivered. In fact, it's unprecedented and has really turned some heads within our various groups.

Was this a reflection of mistakes made by our team working on the software side? Absolutely not. This story really has much more to do with what our small hardware verification team did *right*. What we experienced was an interesting example of how introducing new verification methodologies into a real-world design environment can improve overall productivity and process management.

On the project, we employed a metric-driven process-based approach for the functional verification of our FPGA. I'm eager to tell you how it worked because if you want to save some time, reduce risk, and further improve your verification process, you may want to apply some of the lessons we learned.

The Task Looked to be Complex

Ethernet IP service processors and systems tend to have wide application-specific functionality. Ours was no exception. We were dealing with a highly complex and custom FPGA that required a high level of verification – one that would be a central and key component of the full networking processor system. Our PISA FPGA is a complex block-level packet processor designed to deliver application-level intelligence for L4–L7 switching applications. It is a key component of a board that goes on top of the system supervisor in a regular routing engine.

My team was put on the project midstream and it quickly became clear to me a tough task would lie ahead. One of the most challenging parts of it would come from the verification of the FPGA and its many interfaces. Granted, we'd be leveraging technology from a prior project, but still, our custom design demands were significant and put a tremendous strain on limited verification resources.

Recognizing how crucial it would be for our small verification team to ensure the design is functionally correct well in advance of the debugging cycle, I felt compelled to establish an aggressive schedule, and thus, an effective strategy for limiting the project's risk. Our concern was not only risk to the specific design quality, but also schedule risk and the team's ability to get up to speed on the new solutions we'd be employing.

Looking Back

Our objective was clear, but the same could not be said for our execution strategy. I say this because while the team had a good deal of experience in FPGA design, much of the verification methodology to be introduced would pose new challenges. The team was well acquainted with traditional Verilog-directed test verification, but we needed a solution that introduced a whole new level of automation. We decided to give Cadence's Specman Elite a look.

Our first step involved helping the members of the team get acquainted with the new testbench automation solution. We did this by applying the software to an earlier design to gain some practical verification experience. This design was one that had already achieved a good deal of stability, so it provided a good training ground for mastering the solution.

We found that Specman delivered a very comprehensive environment for verification. It introduced automation levels we had not seen before including functional testing, coverage analysis, and much more. We discovered it is an extremely powerful tool largely because it randomized the tests, created verification scenarios, and sequences automatically, and it was very thorough in its ability to find bugs in the tougher areas. It also leveraged a well thought-out methodology for full verification closure.

With a combination of in-house resources and some additional talent, we went to several training courses to master the tools. While we didn't catch any new bugs on our initial preverified design we did gain the valuable hands-on experience we were after. Mission accomplished.

With our newfound solution, we felt ready to apply the knowledge we had gained on the PISA project ahead. Our training team knew how to perform powerful functional verification on the design. However we still needed a better way to manage the complex project step by step.

Discovering Project Predictability

From the onset of the project we had two main objectives. First, get the verification environment setup correctly. Second, take the necessary steps toward implementing actual management software that uses a metric-driven approach one that would be able to manage, track, and measure progress of our initial plan to full verification closure. We had heard about a solution called Verification Manager that was supposed to work very closely with the testbench automation solution, so we evaluated it.

On past projects, our progress reports were very informal and manual. In fact, they were really just estimates, done for the most part on spreadsheets, Word documents, or status updates sent via e-mail. These approaches were no longer acceptable. We needed a way to be accurate and pinpoint areas that needed more resources or verification cycles. Verification management software offers better resource utilization, a more predictable process, and measuring capabilities for achieving closure in an off-the-shelf software package.

In other words, we came to realize that the ad hoc management we had depended on in the past was only adding a greater degree of risk. What if we got to a point very far down in the verification cycle and bumped up against a bug that required specification changes? We'd be in a very tight spot with limited ability to react, unless we could let the entire schedule slip. On the PISA project, we couldn't have that happen. We needed a good snapshot of where we were on the project at all times, data-driven insights into what the path forward would look like, and insights into if and where we would need to increase or redirect resources.

A Coverage-Driven Approach, a Metric-Driven Environment

The support team we encountered on this project was exceptional. They responded by assigning small teams that were able to continue our training and help us understand the valuable links between the management solutions, testbench results, and our own design's feature set as measured against our verification plan.

The ultimate goal was to have the management solution oversee the process from the get-go. This way as we worked toward completing the established verification objectives, team members could continuously access the progress reports. The tools worked together seamlessly within the simulation environment to give us a complete view of the project.

To get this project-level perspective from the management software we'd call up reports that were easy-to-read HTML files. These reports shed light on the different areas of the feature plan. As the design team manager, I had an interface that gave me a high-level view of the various coverage metrics, which helped me pinpoint the holes. This capability was extremely valuable because it helped me determine how to prioritize tasks and allocate our limited resources accordingly. We found the verification engineers could even do this by themselves, which proved to be a valuable time-saver.

We found that the workflow of this new methodology was relatively straightforward. Basically you launch the management software, select your file, and then read the session. If you choose to, you can look at the main window giving you a complete summary, or you can drill down to look even more closely at each measurement and keep track of the specific coverage metrics.

A New Level of Confidence

In the past, we worked without this level of information, or essentially, without this level of confidence in the accuracy of our verification progress and coverage. Today very little is left to chance or speculation. We have an extremely high level of confidence in the accuracy of the verification process. We operate with much higher confidence not only in the functionality of the device, but in the progress and overall management of the verification flow as well.

We're now in the software QA phase of our project and enjoying a more effective form of information sharing. When we identify a bug, we can use that information relatively quickly to complete fixes much easier and without having to blow out our entire schedule. When we find a strange behavior or occurrence within the design as reported by the software, we try to reproduce it with simulation first, even before debugging in the lab. We can identify these failures with greater detail and report our findings to the entire design team faster.

This is how we're able to stay on schedule, or in this case, actually do better than our schedule. I feel confident in telling you that we have found an invaluable way to reduce our project risk – and that's my job. Perhaps it's your job too.

Developing a Coverage-Driven SoC Methodology

Andreas Dieckmann, Siemens AG
Automation and Drives

Dr. Andreas Dieckmann lives with his family in Nürnberg, Germany. In 1995, after obtaining his MA at the University of Erlangen and his Ph.D. in Electronic Engineering at Technical University of Munich, he began working at Siemens AG. Initially he was involved in board and fault simulation. From 1997, Dr. Dieckmann gained expertise in system simulation and verification of ASICs. Since 2001, he has been in charge of coordinating and leading several verification projects employing simulation with VHDL and Specman "e," formal property and equivalence checking, emulation and prototyping. The case study described here is an extension of the coverage-driven methodology developed by his team for the verification of SoC projects.

Introduction

Methodology is the key to successful verification of complex SoC designs. There are many verification tools, techniques, and languages available today, and many of these can be quite effective if used properly. However, effective usage requires a comprehensive methodology to link together such seemingly disparate approaches as simulation, formal analysis, and prototyping as well as the multiple languages commonly used for verification.

This article describes a methodology developed by our verification team in the Automation and Drives (A&D) group of Siemens AG.

This methodology has evolved over our past few projects and has reached its current form in the verification of two related ASIC SoC designs, each containing about 4M logic gates and numerous small memories totaling about 1 MB of SRAM. These two chips were significantly more complex than previous projects and required a multisite development team, fueling our methodology evolution.

This article provides some background on older projects and describes in detail the coverage-driven methodology in use today. While we viewed the two chips as pilot projects to develop a methodology that could be extended to other projects at Siemens A&D, we know that our verification needs would continue to grow and that our methodology would continue to evolve. Accordingly, we conclude by discussing some likely enhancements for future projects.

Verification Background

Although the two-ASIC project added a number of new requirements and challenges to our verification process, in fact we have been in a process of continual improvement for many years. As was the case for many European design teams, we long ago chose VHDL as our RTL design language due to its early standardization and its superior capabilities (user-defined and enumerated types, package and generate statements, library support, etc.) over original Verilog.

However, our choice of VHDL was also made with verification in mind. Its advanced constructs allowed us to build more sophisticated and more reusable testbenches than was possible with Verilog. Thus, up until 2001, our verification environments were VHDL-centric, with both the RTL design and the majority of the testbench code in VHDL.

2001 saw the next major step in the evolution of our verification process, when we chose the *e* language and the Incisive® Enterprise Specman® Elite testbench automation solution (now available from Cadence). We had found that adding randomization to our VHDL testbenches brought great benefits in terms of finding bugs more

Developing a Coverage-Driven SoC Methodology   287

quickly; Specman Elite's constrained random stimulus generation capabilities made it even easier to thoroughly exercise our designs.

Our adoption of *e* and the constrained random approach led to a related evolution in our methodology: Our verification plans gradually shifted from test focused to feature focused. Figure 1 shows an example of a traditional verification plan that lists the tests to be written for each major functional unit in a chip and tracks the status of test completion. Such plans are often called "test plans" although this term is probably better reserved for physical chip testing.

Functional Unit	Test Name	Spec Written	Test Written	Test Passed
Bus Interface	read_sequence_a	X	X	X
	read_sequence_b	X	X	X
	write_sequence	X		
	r_w_intermixed			
Cache controller	cache_hits	X	X	X
	cache_misses			
	cache_flush	X		
Interrupt FSM	exercise_all_states	X	X	

Figure 1 The Traditional Verification Plan

The problem with this traditional approach is that it requires a precise mapping from functional units to specific tests. That makes sense when the tests are hand-written to test-specific areas of the design. However, constrained random stimulus generation may exercise many areas of the design at once and can run as long as the user chooses, so the notion of an individual test is no longer a useful one.

This observation raises the question of how verification engineers can tell what a constrained random test run is actually exercising.

The answer is that some sort of coverage metric is needed in order to provide a quantitative measure of verification effectiveness. With such a metric in place, we can say that a test run verified all areas that it covered and, ideally, we can combine the results from all test runs to get an overall view of coverage.

In terms of specific coverage metrics, we have made extensive use of functional coverage but minimal use of code coverage on our projects. We have found that tracking functional coverage points provides a much better measure for determining what each test run accomplished and where we are in terms of overall verification completeness.

As we began the project with the two 4M-gate SoCs, we decided to add the Cadence Incisive Enterprise Manager verification management with process automation solution to our arsenal of tools. Enterprise Manager provides a mechanism to capture the features in our design and – in concert with Specman Elite – reports functional coverage results against these features. Figure 2 shows a screen shot of one such report in HTML format.

Section	Rel. Grade	Grade	Completion	Goal	Weight	At Least	Valid	Sampled	Items	Planned
2.1.1.3.4 AHB Multilayer Medium Speed MSML AHB			100%	100	1	1	1889	1730	120	120
2.1.1.3.4.1 MSML AHB Bus Protokoll (specman)	91%	91%	100%	100	1	1	1633	1487	101	101
2.1.1.3.4.2 Verbindungsmatrix MSML AHB (specman)	98%	98%	100%	100	1	1	50	49	1	1
2.1.1.3.4.3 MSML AHB Arbitrierung	92%	92%	100%	100	1	1	186	180	6	6
2.1.1.3.4.4 MSML AHB Burst Breaker (specman)	100%	100%	100%	100	1	1	5	5	5	5
2.1.1.3.4.5 MSML AHB SLAVE2LITE	100%	100%	100%	100	1	1	15	9	7	7

Figure 2 A Modern Verification Plan (vPlan)

The combination of tools and techniques in our current verification environment enables a true coverage-driven methodology. As described in the next section, we put a great deal of effort into defining detailed, corner-case features in our verification plans and in specifying e functional coverage points to track the exercise of these features. Thus, we continue constrained random test runs when features remain uncovered, and use the composite functional coverage results as a key factor is determining when to "tape out" (release the netlist to our ASIC vendor).

Current Verification Methodology

Since a methodology is an abstract concept that's hard to visualize, we tend to think in terms of the verification flow enabled and supported by the methodology. Figure 3 provides an overview of our flow, starting with a functional specification for an SoC, employing multiple methods to thoroughly exercise the design, and reporting results against the verification plan (vPlan) defined with the help of Enterprise Manager.

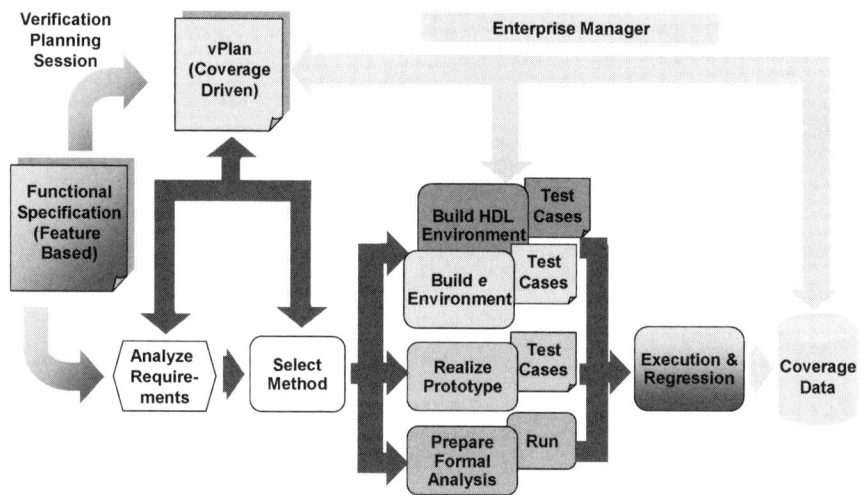

Figure 3 Coverage-Driven Verification flow

While the SoC architects complete the functional specification, our verification process starts with a series of planning sessions for the 20–40 modules in a typical chip. These meetings usually involve 5–6 people, sometimes more for large modules, and include design engineers, verification engineers, and the specification writers. Their job in each session is to develop a detailed feature list for each module and capture it online in a vPlan so that coverage results can be automatically tracked during the verification process. The session also serves as a detailed review of the functional specification, further motivating the design engineers and specification writers to attend.

In order to make the tracking as precise as possible, each feature has to be related to at least one functional coverage point. Thus, the features are quite fine grained and often reflect important corner-case conditions in the design. Because of this, designer involvement in the planning process is critical. Verification engineers usually don't know enough about the details of the RTL implementation to understand all the critical corner cases.

We use four major approaches to verifying the design. For exercising the features in the vPlan, we strongly rely on *e*-based verification environments for individual modules and for the chip as a whole. We develop the necessary transactors to generate constrained random stimulus and check results, while also writing the functional coverage code to monitor each vPlan feature. The testcases in our *e* environments are almost entirely constrained random test runs; we try to minimize the need for hand-written directed tests.

We also use HDL-based (Verilog or VHDL) simulation environments to verify some specific modules. For example, we sometimes have IP blocks or models that have their own HDL-based testbenches. In such cases, we may make use of code coverage metrics in order to assess verification thoroughness since we don't have any functional coverage points in the HDL testbenches.

At the submodule level, we sometimes make use of Incisive Formal Verifier (IFV) and its formal analysis to complement the simulation-based environments. As part of our verification planning process, we identify portions of the design for which we can specify assertions that cover 100% of the interesting behavior and then use IFV to target these assertions for proof. We write our assertions using the VHDL "flavor" of the PSL.

Examples of our formal analysis usage include bus-multiplexing structures and memory correction algorithms. While we targeted capturing 100% of the intended behavior with properties, in practice there is no way to know this for sure. However, all formally verified submodules are also tested as part of the module's environment, and so we use formal analysis and simulation as complementary, rather than contrasting, verification approaches.

At the full-chip level, we develop an *e*-based environment that focuses on verifying the proper interconnection and integration of the modules, not on verifying the functionality of the modules themselves. The chip-level tests are therefore fairly simple, and we do not put a lot of effort into tracking coverage metrics at this level. All of our *e*-based tests, for both modules and the full chip, are run using the regression–automation capabilities of Enterprise Manager.

Most of our current SoCs contain at least one embedded processor core, and in such cases we take a further step for full-chip simulation. We develop an HDL-based environment in which we run self-checking C testcases directly on the embedded core. This ensures that the core can access all the functional modules and put them into operation.

Since the speed of chip-level simulation limits the length of the tests that we can run, we also make use of an FPGA-based prototype to run real application code. We find that it is hard to correlate application-level testing with specific features, so we do not currently have a method to gather coverage data from the prototype and combine it with simulation results. We do write some prototype tests to exercise specific behaviors, such as taking a timer through its full count-down range; validating that this actually occurs can be viewed as another form of coverage.

We do collect the functional coverage data from the module-level *e* environments and use Enterprise Manager to report the coverage metrics against the features in the vPlan. This provides a coverage-closure loop that lies at the heart of our coverage-driven verification methodology. As we will discuss in the summary, we hope to extend this loop to include all aspects of our verification flow.

Coverage and Checking

One common area of confusion for new adopters of coverage-driven verification is the role played by coverage metrics and the role played by checkers. The distinction is actually rather simple: Coverage tells whether something happened while checking tells whether something happened *correctly* (per the functional specification).

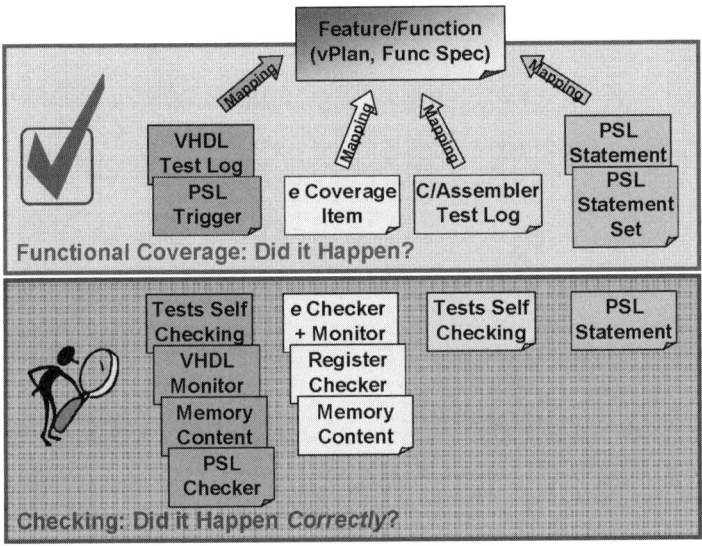

Figure 4 Coverage Metrics and Checks

Figure 4 shows some important components of the overall verification environment and how they contribute to these two areas. As previously mentioned, we make minimal use of code coverage so our coverage focus is on functionality. In HDL testing, we sometimes include dedicated testbench code to check for certain desired behavior and log when it occurs. This is really a type of coverage, albeit a more informal ad hoc method.

Developing a Coverage-Driven SoC Methodology

In addition, the PSL properties can be monitored in simulation to detect when they were triggered. For example, consider the following PSL statement:

```
property p is always ( REQ -> next
ACK) @rising_edge(CLK);
```

This property specifies that, whenever REQ is asserted, ACK must be asserted in the next cycle. This property is considered triggered whenever REQ is asserted on a rising clock, so in some sense this property has been covered.

When we write C and assembly language testcases to run on the FPGA-based prototype, much as we do for HDL simulations, we sometimes include code to check for behavior and log it. Again, this can be regarded as another form of coverage. Finally, in the results from formal analysis, we track which properties were proven as well as which properties were triggered even if formal was unable to complete a proof.

As mentioned previously, we have a clear mapping between vPlan features and *e* functional coverage points. The mapping from HDL, formal, and prototype coverage information is not truly automated in our current verification environment, and so some of the links shown at the top of Figure 4 are more theoretical than actual.

The bottom of this figure shows the checking components. Our simulation and prototype tests are normally self-checking so that a definitive answer is provided in terms of correct behavior. We also use a number of verification components in the simulation environments to check for protocol compliance, correct register and memory contents, and passing assertion properties. These same properties are also used in formal analysis, which reports either a proof of correctness or a bug.

Results and Futures

We have been very pleased with the results of using the coverage-driven verification methodology on our two latest SoC projects. One measure of this is our consistent discovery of bugs throughout the verification process, as shown by the example in Figure 5.

Authors note to Figure 5: The legend entries directly correspond to the level of each data band in the associated chart.

The nature of constrained random stimulus generation means that we can continually run additional tests, experiment with different seeds to vary the random behavior, or tweak biases to produce a better mix of stimulus (such as the ratio of reads and writes on a bus) as long as we keep finding bugs. Observing the bug-discovery rate and tracking coverage metrics are both important contributors to the tape-out decision.

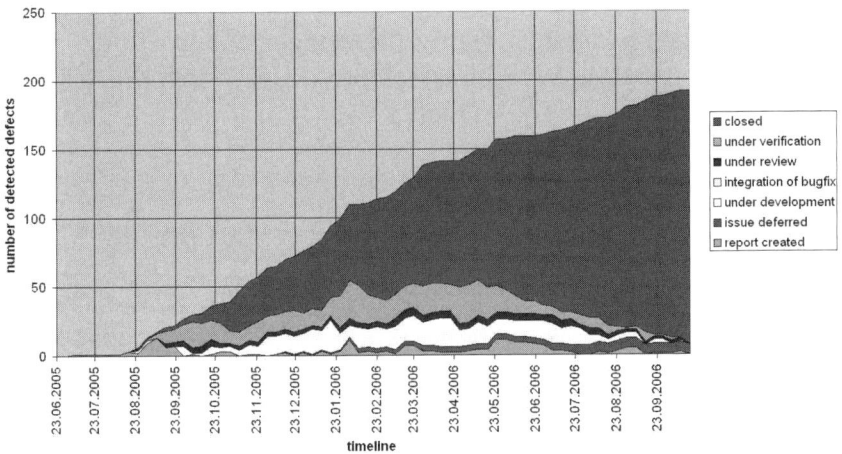

Figure 5 Defects Detected Over Time

We embraced the coverage-driven methodology with three goals in mind:

- Better verification schedule predictability
- Design quality guarantees
- Improved reuse

The faster and more thorough bug discovery achieved on the two-ASIC project satisfied our first two goals. We also met our third goal; many components of our verification environment were reusable from the module level to the full-chip level, and many will be reusable on future projects as well. We followed the Cadence *e* Reuse Methodology (*e*RM) in developing about two dozen *e* Verification Components (*e*VCs) for various interfaces and functions inside these two chips. In addition, we used commercial eVCs for standard interfaces such as PCI and USB.

As happy as we are with our results so far, we have a number of ideas for improving and extending our methodology for future projects. As mentioned previously, we would like to find more effective ways to bring in coverage metrics from HDL simulation, formal analysis, and prototype testing to automatically combine them with the *e* functional coverage results. Both Specman Elite and Enterprise Manager have capabilities for importing coverage data that we have not yet tried.

We have also not yet taken advantage of all the hierarchical planning features of Enterprise Manager, but we plan to do so in order to combine results from our module-level and chip-level *e* verification environments more easily. Since our verification focus is different at these two levels, it makes sense to look at a single view rather than separate results. We believe that this will lead us to a true plan-to-closure verification methodology, in which every step of our process will be correlated back to a unified vPlan.

As future projects get more complex, we expect that significantly more verification cycles will be needed. As our regression tests get longer, we will likely need to run on server farms rather than only a few machines, a capability supported by Enterprise Manager. Also, we will probably want to use the test-ranking features of Specman Elite to automatically select subsets of our full regression suites for rapid verification of RTL changes.

In summary, the last 5 years or so have been a period of rapid evolution for our verification team at Siemens A&D. We have moved from VHDL-based directed tests into a constrained random, coverage-driven approach complemented by formal analysis and prototyping, all tied together by a comprehensive methodology. Given our planned enhancements, we are confident that our verification methodology will continue to evolve to keep pace with our future project demands.

From Panic-Driven to Plan-Driven Verification Managing the Transition

Susan Peterson and Paul Carzola, Cadence Design Systems

Susan Peterson has been trying to escape from the EDA industry for the past 20 years, where she has spent her time listening to customers and trying to help them to solve their critical problems in various sales and marketing roles. Prior to that, she was a practicing engineer, and earned her MBA from the University of Denver. She enjoys everything outside from her home and office in Colorado.

Paul Carzola is a Senior Consulting Engineer for Verification at Cadence. He received a Bachelor of Science Degree in Computer Engineering at Florida Atlantic University in 1995. Since then, Paul has spent the last 10 years in Functional Verification and the pursuit to finding effective and powerful methods to verification while making it easier and enjoyable to apply. For the past 5 years, he has served in a consulting role in the area of functional verification methodology and has seen first hand the power of a coverage-driven approach.

Begin with the end in mind. Look before you leap. Eat your vegetables.

Sounds so right! Similarly, taking a structured, measurable approach to making sure your design works right the first time through just makes good sense. So why is there so much foot-dragging when it comes to implementing a metric-driven approach to verification?

"I know there's a train wreck coming, I'm just not sure where to look for it," says one project manager for a 10 million+ gate complex SoC. "We have two junior engineers pumping out directed tests just as fast as they can over in the corner there – but I'm concerned that this time it just won't be enough. If this tape-out works, it'll be pure dumb luck." With millions of dollars riding on a respin and shrinking market windows, it's no wonder that project managers worldwide are losing sleep.

They know there's a better way. So why don't they take it? In general, because it just seems like too far to go from here (two junior engineers in the corner) to there (a metric-driven verification plan with feedback loops all along the way to keep you on track).

Nobody likes change. But change we must if we want to regain some peace of mind on the way to tape. So take a deep breath as you consider the following steps to implementing the changes you know you need.

Take Stock
Like any good problem, defining the "as is" state is a great place to begin. How would you rate your verification process and where are areas for improvement? Ask yourself the following questions:

- *How do you make the call that you're done with verification?* If the answer is "when we're out of time," don't worry – you're not alone. In fact, this is overwhelmingly the most frequent answer, and the best indicator that a change in your verification methodology is long past due.

- *Are you afraid of the bugs that are hiding in an untested area?* Do you feel overwhelmed with the amount of tests that have to be written? Maybe all of your directed tests have passed – but how do you know you've tested every path that 13-year old user might think up? Without a closed loop system that ties your metrics back to the original intent, your verification is incomplete – and is another good indicator that you're going to need to upgrade your verification methodology to shut-up that nagging little voice inside of you that keeps questioning whether you're really done.

- *What are you measuring now?* How is it tied back into the architect's original intent? Maybe you've set 100% code coverage as a metric – which is a great way to measure whether all of your code has been tested. But how do you know whether it does what the system architect envisioned it would do? You're going to need to branch out if you want measurements that reflect on whether your design will work as architected.

- *What verification expertise do you currently have on your team?* Do you need more? Are your engineers trained up with the latest technology and methodologies? Face it – you're going to need to make an investment in people and the tools they need to solve this.

Invest in Verification

If you're like most people, you'll probably say that you only spend about 20–30% of your development process in design. After all, you've invested in HDLs, Synthesis, Timing analysis, and all the training you can afford – it's no wonder you have such a nice balance between the time you spend in design, the tools you have to do the job right and the talented people you need to get it done.

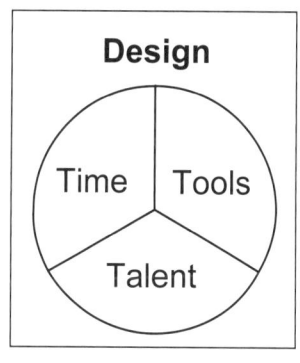

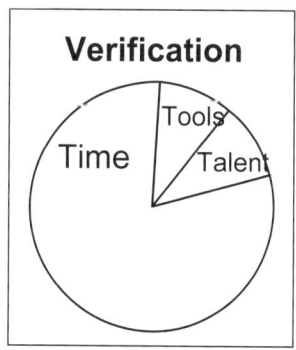

On the flip-side, most teams would agree that they spend about 70–80% of their development process in verification, and never really feel confident in their results. Is it really any wonder? For the most part they're using languages that were built for design. Verilog, C, or Perl were never meant for verification and so much time is spent creating an infrastructure rather than tackling the true problem, the chip itself. Without a good methodology, verification is usually an ad hoc, "shoot from the hip" process, and is frequently viewed as an unglorified role filled by the most junior people on the team who haven't earned the right to be designers.

Thus, the time that verification takes is often grossly out of balance due to lack of investment in tools and talent. An investment in verification talent and tools can reap huge rewards in shortening your overall development process and giving you confidence that your design will work as intended the first time through.

Reuse It or Lose It
If you wanted to build a car, you wouldn't reinvent the wheel. Similarly, from project to project, person to person, protocol to protocol – I'm betting that there are large blocks of your design and verification environment that could be reused if:

- Your teams followed a consistent methodology for creating the blocks
- Somebody took the time to make everyone aware of the IP and how to use it
- You looked outside your company to find verification IP and expertise for commonly used protocols

So get organized.

Begin With the End in Mind
What's the biggest hurdle in your critical path? Don't know? Well maybe that tells you something. Wasn't it Deming that said, "If you can't measure it, you can't make it better?" So at the risk of being too pedantic, you really do need to begin with the end in mind.

We've all heard it before: "You need to start with a plan". And most people would say they are using one or several plans. But after tape-out how often does your plan reflect what you actually measured from the device? Typically, engineers see creating a plan as a time-consuming chore with little value. All plans begin with good intentions but inevitably other issues arise during the verification process and updating the plan is not viewed as important. Panic sets in.

It's important to give value to the verification plan and make it a regular part of the verification process. Have regular meeting and reviews on where the project is with respect the verification plan. This will allow you to redirect efforts to attack weak areas or make adjustments to the plan to accommodate new direction.

You should consider investing in tools that allow the plan to be executable and directly tied to the results. This will allow users to instantaneously see the plan with their perspective in mind and make any necessary course corrections.

Learn from the Best
So, you looked inside your organization and didn't find anybody who knew that much more than you do about verification? Maybe it's time to look outside. Sure, you can train the people you have, but expect push-back from your senior engineers when those new engineers you just trained come back from training bursting with new, and sometimes foreign, ideas. And then there's the learning curve.

Instead, consider looking outside your team and even your organization for some new, experienced blood. You may be surprised to find that there are many consulting companies that offer decades of experience leveraged by the newest tools and methodologies ready to help springboard you from here to there. They offer great return on your investment and leave your team re-energized and more valuable in their wake. All you have to do is ask.

Sometimes we have to listen to the children, as with this quote from Lewis Carroll's "Alice in Wonderland":

One day Alice came to a fork in the road and saw a Cheshire cat in a tree. "Which road do I take?" she asked. "Where do you want to go?" was his response. "I don't know," Alice answered. "Then," said the cat, "it doesn't matter."

As long as you're embarking on the road to better verification, make it matter. Take stock. Start with a plan and make it an active component. Invest in tools and IP. Find an expert to help you. But by all means, take the first step.

Verification of a Next-Generation Single-Chip Analog TV and Digital TV ASIC

YJ Patil, Genesis Microchips and Dean D'Mello, Cadence Design Systems

YJ Patil is a senior verification engineer at Genesis Microchip, where he is responsible for managing the verification of Digital Television (DTV) controller ASICs. Prior to Genesis, Mr. Patil was a verification engineer at several technology leaders including ATI, Silicon Access Networks, and Philips Semiconductors. He was a board designer at Tektronix. Mr. Patil holds an M.S in Software Systems from BITS Pilani, India and B.Eng. in Electronics and Communication from Gulbarga University, India.

Dean D'Mello is a Solutions Architect at Cadence Design Systems. He works closely with key customers worldwide to deploy advanced verification technologies, and with R&D to plan, develop, and introduce new methodologies and products. Prior to Cadence, Mr. D'Mello held ASIC design and verification roles at LSI Logic, Cogency Semiconductor, and Celestica, and product and test engineering roles at IBM. Dean holds a Masters of Applied Science (MASc) in Electrical and Computer Engineering from the University of Toronto, Canada.

Abstract

Consumer demand for entertainment products brings to verification engineering teams the challenge of verifying designs which integrate functions from multiple previous-generation products with new features. Effective reuse of verification code combined with judicious adoption of new verification technology is needed to achieve the productivity required to meet project schedules. From the perspective of a verification lead, this paper presents the approach taken by a verification team working from multiple sites to verify a multimillion-gate ASIC which implements an Analog and Digital TV solution. Choices made by the team to reuse existing code, build new verification components, and adopt new technologies to meet the needs of the project will be described, along with results achieved on the project, for which verification was completed on schedule in half the time and with half the engineering resources required to verify the previous ASIC.

Introduction

High-Definition TV (HDTV) offers consumers a rich video and audio experience. Consumers demand quality products at competetive prices, and FCC mandates require consumer-electronics manufacturers and broadcasters to keep up. This brings the challenge of integrating existing Analog TV (ATV) solutions with Digital TV (DTV) to facilitate a smooth transition and address the needs of a variety of users.

These market requirements drive architectural changes in ASIC designs. A cost-efficient solution requires tight integration of ATV and DTV solutions into an ASIC. The integration needs to address the grouping of functions, sharing of resources and adding new interfaces. For example, new interfaces need to be added to address the higher bandwidth requirements of the combined solution. Verification challenges lie in dealing with the complex design, the new interfaces, modified blocks, and managing the huge set of test suites. Tight schedules and limited engineering and computing resources are the constraints to the problem domain that need to be addressed.

Previous approaches by verification teams to these problems have been to add more engineering and computing resources, and create custom tools to measure and track verification progress. This case study of the verification of a next-generation single-chip ATV and DTV ASIC with numerous analog interfaces and several embedded processors (DSP and RISC) describes how some of these verification challenges were dealt with on a recent project. Choices made by the team to reuse existing code, build new verification components, and adopt new technologies to meet the needs of the project will be discussed in this chapter. The combination of verification strategy, tools, and methodology enabled the completion of verification in half the time and with half the engineering resources required for the verification of previous ASIC, enabling delivery of the final product within the required market window. Some of the limitations of the approach and opportunities for further improvement are also discussed.

This Chapter is organized as follows: The remainder of this section introduces the DUV and its verification challenges. Following that, we describe the key enablers of results achieved namely strategy, verification planning, and verification environment implementation. Finally, we summarize the results achieved and identify areas for improvement on future projects.

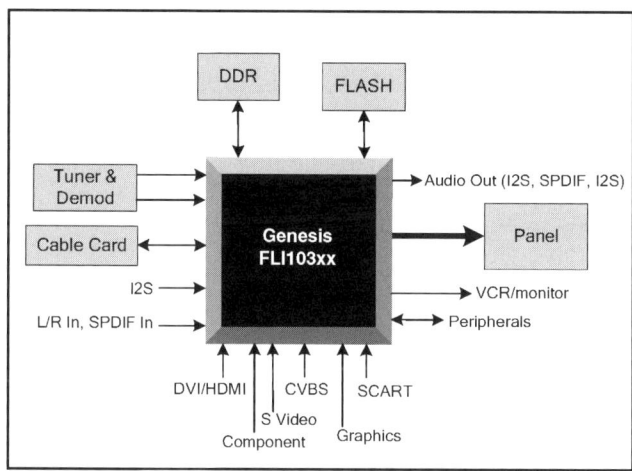

Figure 1 The Design-Genesis FLI103XX

The Design

The Genesis PūrVIEW HD™ 300 Series IC (FLI103xx) is a single-chip TV solution for products requiring superior video quality in the analog and/or digital TV for ATSC, DVB, NTSC, OpenCable, and PAL markets. This solution includes a single channel HD MPEG2 decoder, flexible analog front end with an integrated Faroudja® 3D Video Decoder, high-performance industry standard 32-bit MIPS

4Kec™ processor (250 MIPS), multistandard analog audio decoder, digital audio decoder and post processor, three programmable multimedia processing engines (MPE), advanced 2D graphics engine, integrated HDMI/DVI receivers with HDCP support, unified DDR memory controller and a very flexible and unique Video eXpansion Interface (VXI) providing glueless connectivity to Genesis video coprocessors, or a customer's proprietary video processing chip.

The solution includes next generation Faroudja DCDi Cinema® video format conversion, video enhancement, and noise reduction. The level of video quality that could previously only be seen on an exclusive Faroudja Home Theater System is now available in a single-chip solution.

The interfaces of the single chip ATV–DTV are shown in Figure 1.

Verification Challenges

Some of the factors that contributed to verification complexity on the project were the number of configurable registers (programmability), the number of interfaces, and the number of data paths (Table 1).

Table 1 Problem Domain Description

Indicator	*Previous chip*	*ATV–DTV chip*	*Change*
Number of registers	1600	2600	+ 63%
Number of external interfaces	7	12	5 NEW interfaces
Regression size	500	1100	+120%
Number of DUVs	TBD	24	TBD
Subsystems	DTV	DTV, ATV, and CPU	Two more subsystems
Logic gates	Not disclosed	Not disclosed	DOUBLE

Table 1 enumerates some of the contributing factors to the verification complexity, and compares them to the previous project (DTV ASIC) undertaken by this verification team. As shown in the table key factors have doubled.

The generic challenges which arise from integration are Architectural changes in various areas, Addition of new internal buses, Module-level verification, Data paths and integration verification (chip level), and Management of verification process and data.

Addition of New Internal Buses

Integration of two different designs leads to the optimization of resources in the architecture. ATV/DTV design integration resulted in shared memory controller. The memory controller now needs to serve two clients which demand higher bandwidth compared to the previous designs. This required the creation of a new low-speed register configuration bus to offload traffic from the main system bus. Introduction of the low-speed bus (which connects to almost every block) required reverification/regression of those blocks and changing the configuration sequences. The new bus-architecture also required rearrangement of the address map and hence reverification of address decoding logic.

Architectural Changes in Various Areas

Architectural changes are driven by two major factors. The first is the integration of two designs and the second is the request for new features by customers. The color coded diagram (Figure 2) shows the modifications to the architecture.

The changes due to the integration included:

- Addition of a new low-speed bus
- Deletion of the one of embedded control processors and one set of standard peripheral interfaces
- Addition of a new interface block between the two designs
- Modification of the memory controller
- Addition of the DTV subsystem to the set of video sources processed by the video enhancement engine

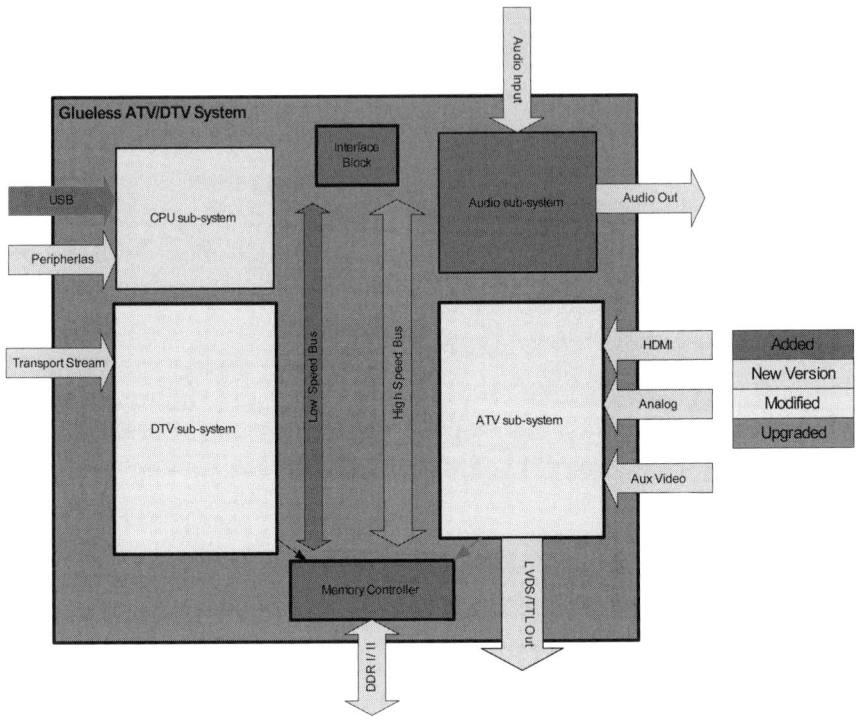

Figure 2 Architectural Changes

The new features driven by customer demand included adding the ability to connect a USB device to view pictures from cameras on the TV screen. The USB connection could also be used to do firmware upgrades and provide an interface for On-Screen-Display (OSD). Also added was high-quality audio, with multiple channel audio inputs and outputs, resulted in the integration of the whole audio subsystem from different designs.

Architectural changes which will lead to the addition/deletion/ modification of blocks affect the volume of verification jobs. These changes also define the verification problem in an interesting way, wherein you need to maintain the integrity of the old design while verifying new features incrementally.

Module-Level Verification

Although this project was primarily an integration of two existing products, some new modules added to the design needed to be verified in the context of the current design. The integration and various modes of modules needed to be validated both by stand-alone module qualification verification and integration testing. New modules added to this design included an Audio subsystem and a USB–AHB bridge. Since most old modules were changed, the newer versions of these modules needed to be verified. For example, the control processor CPU and video decoder cores were replaced with new and advanced versions.

Data Paths and Integration Verification

When two systems are integrated, integration verification is critical and must address several questions to determine its success. The verification environment should reuse the components built around the subsystems. If these two subsystems come from two different verification worlds then the reuse, and hence chip-level verification, becomes challenging. It is also required to identify what gets verified at the chip-level and what gets verified at levels lower in the hierarchy. The key technique is to abstract away from the full detail of the design, just retaining the sufficient features to prove there are no intersubsystem interconnection issues. For example, for the feature set relating to the playing of digital video, top-level verification was used for the playing of digital video from different sources (e.g., MPEG, HDMI), while block-level verification focused on error handling in all different modes from a specific source.

Management of Verification Process and Data

The short schedule cycle of the consumer world and tight engineering resources do not offer much flexibility in finding solutions. The cost of missing schedule deadlines can force teams to terminate the verification effort prematurely. Thus, ASIC design quality can become a function of verification schedule, rather than the verification metrics, making it critical to track at a detailed level which features are verified and which remain unverified, to make the good tradeoffs between quality and schedule.

The large number of DUVs and a verification team distributed between Silicon Valley, Toronto, and India necessitated standardizing the regression infrastructure. Scripts and progress-tracking schemes across the project, equipping every engineer with the ability to launch track and analyze their own regressions, while providing the verification lead with the capability to assess the progress of each of the verification subprojects.

Key Enablers of the Solution

Having defined the verification challenges in the previous section, we now turn our attention to the key enablers of the verification solution that resulted in on-time completion of the verification of the DUV. This section begins with a description of the strategy adopted for verifying the device, followed by discussion of the verification planning and management methodology and solution, and some elements of the verification environment implementation, namely register verification and the considerations and choices made to reuse or build new verification components and environments.

Verification Strategy
From the perspective of a verification lead, a sound strategy needs to be formulated before teams in multiple sites begin work on a project of this magnitude.

The verification strategy followed was to scope the problem, assign the right resources, and devise a complete plan. The development process tends to begin with a high-level understanding of the product requirements and a first-cut at schedule, with more detail added as the architects and designers begin their work. Verification strategy needs to accommodate this development process. The subsections that follow describe the process followed to devise a strategy that quantified the problem at high level, allocated resources accordingly, and was flexible enough to be applied iteratively as changes were made to requirements or designs.

Scope the Problem
Based on the architectural changes (which were driven by integration and customer needs), a delete/add/modify/no change (X/A/M/NC) matrix was identified. Table 2 shows an example of the analysis.

Identification of the number of DUVs was done as the next step to identify what needs to be verified and in which hierarchy (block subsystem/chip level).

Table 2 Verification Resource Mapped to Blocks

Block Name	Arch Change	Verification Resources
Memory controller	Add	Full-time
DMA controller	No change	Shared
Peripherals	Modify	Shared
CPU bus	Add	Full-time
High-speed bus	Modify	Shared
Video memory arbiter	Modify	Full-time
186 Core	Delete	None

The X/A/M/NC analysis matrix helped to get the assessment of the kind of work involved. There were options such as assigning one dedicated resource to the block which went through modifications. There were generic changes. For example, the "new bus" changed the interface to all blocks. Twenty-four DUVs were identified; each of these DUVs went through either minor modifications or major modifications to be able to integrate in new architecture.

Resource Allocation

Exhaustive block-level verification was required to thoroughly verify all blocks which were added or went through major modifications. The block owner would create the block-level verification environment, interface verification components, monitors, scoreboard, etc. The block owner would also write the verification plan and define coverage necessary to measure compliance to the plan and execute it to verify the planned features

As described in Table 3, the peripheral bus and all those blocks mapped to this bus were assigned to a single owner. The owner is responsible for bus e-Verification Component (eVC) with master and slave agents. The owner will go through modifications for all the slave block environments and ensure that changes are intact.

This resulted in a single owner being used efficiently to deal with several modified blocks.

Table 3 Verification Resources and Tasks

Resource	DUV	Additional Responsibilities
Engineer-1	Block-M1, M2, and A1	
Engineer-2	Block-M3 and M4	Low-speed bus
Engineer-3	Block-M5, M6, IP1	Mentor for newcomers
Engineer-4	Block-A2, M1, chip-level	
Engineer-5	Peripherals and Chip-level misc logic	Script and regression
Engineer-6	Block-M7, chip-level	Interrupts
Engineer-7	Block-M8, IP1, IP2	
Engineer-8	CPU subsystem, DTV SS	Chip level
Engineer-9	ATV SS, chip-level	
Engineer-10	chip-level, progress tracking	Back-up and debug support

The chip was divided into two major subsystems. One owner for each subsystem ensured all the changes in that level were intact and the subsystem verification plan was tracked to closure. The Chip-level/System-level owner focused on the integration of the subsystem and verified the interconnectivity and data paths derived from user scenarios. Each block owner owned one of the relevant data paths at the chip level.

The compute farm used for simulations included a cluster with 12 machines at one location and another cluster with ten machines at different location. vManager and LSF were used to dispatch the regression jobs.

There were two major usage scenarios for simulation licenses, at least one per engineer during development and enough to get the regression throughput in the later stages (peak load). There was a recommendation on the licenses but in the end the team had more than 2X licenses available than the required amount (around 15 licenses). To manage regressions there was one vManager license per site, but when several engineers needed to do debugging concurrently, the license became a bottleneck and one vManager license per engineer was obtained.

Verification Planning and Managements
The high-level strategy described in the previous section provided a general game-plan for verification activity. The need to manage a large set of functional requirements verified using more than 20 environments across multiple sites and tracked using coverage-driven verification methodology led the team to adopt vManager, a tool from Cadence that automates and assists processes in functional verification. This section describes how vManager was used on the project.

Executable Verification Plan
A verification plan was created for each of the DUVs. The primary purpose of this plan was to identify all features that needed to be verified. This assisted with resource allocation as well as with the tracking of verification progress using vManager.

The plan used in the verification process is termed "executable" because although it is written in natural language using a word-processor, the use of special paragraph formats and export as XML allowed reading of the plan into vManager, and annotation of coverage metrics to each feature, providing a hierarchical feature-based view of progress as measured by coverage collected in simulations. The use of natural language to create the plan allows stakeholders from disciplines outside verification (e.g., Architects, Designers, and Software engineers) to participate in the planning process.

A snap-shot of section of vPlan is Figure 3. Note the hierarchical list of features to be verified, short description of each feature, and identification of coverage groups that measure how well the feature is verified in simulation.

1 Functional Requirements

1.1 Reset and Initialization

   Details: This section will include the reset and initialization features for the full-chip.

1.1.1 Register Accesses

   Details: Read and Write accesses to all registers
   Cover group: vr_ad_reg.reg_access

1.2 Interfaces

   Details: The major internal and external interfaces need to test for their connectivity and functionality at top-level. The following group summarizes all those interfaces. They would be implicitly tested by data flow/control path testing. On each of the interfaces there would set of assertions for ensuring protocol adherence and also coverage monitor is intuitive and hence are placed on interfaces.

1.2.1 Inter-block Interfaces

   Details: Major internal buses that are important from a full-chip perspective are: Pbus, DMA cfg, Dbus, Comm-bus, Generic Control Interface.

1.2.1.1 P-bus Interface

   Cover group: cm_pbus_bus_monitor_u.transaction_end.rw
   Cover group: cm_pbus_bus_monitor_u.transaction_end.addr
   Cover group: cm_pbus_bus_monitor_u.transaction_end.cross__rw_addr

Figure 3 Executable vPlan screen-shot

Regression Management and Progress Tracking

The integration of two designs into a single chip brings with it a lot of tests from the previous development that all need to work in a new, bigger design. Regression of these tests was the starting point. As the project progressed, a regression was run regularly to ensure that the X/A/M/NC features were not breaking the unmodified portion of the design. One of the overall progress indicators was the passing percentage of the regression.

Because of the previous designs inherited and the addition of new tests at the block level, subsystem level and chip level, the regression size became big. The peak time regression size was 1200 tests, each test running for an average for ½ hour. To measure the overall progress, the numbers that needed to be tracked were how much of the verification plan was covered. The goal was to achieve 100% functional coverage and 100% code coverage and the exceptions had to be documented. The use of coverage-driven methodology meant that results of large numbers of simulations would need to be analyzed often by each engineer.

Execution of the verification tasks described above required introduction of a verification management methodology given the large number of DUVs, team members in three sites and the need for the verification lead to track progress of all the verification sub-projects. The main benefits sought from the verification management methodology were:

- All engineers are able to launch and analyze their own regression without having to create and maintain custom scripts
- A standard view of simulation failures from each DUV is available to allow quick analysis to determine how many unique failure signatures were observed, and select a simulation for debug
- Progress reports which summarize coverage results from simulations by DUV feature are available to track progress with respect to the verification plan and identify uncovered areas

vManager is a product from Cadence that automates and assists tasks and processes in functional verification, enabling verification teams to deploy a verification management methodology to achieve the benefits described above.

vManager provides:

- A regression runner infrastructure that provides a standard and simple format for describing simulation sessions, dispatching them to a compute farm using LSF, and tracking and controlling them from a GUI
- An environment for interactive analysis of large number of simulation failures, with the ability to group failures with

similar characteristics, identifies simulations that exhibit each failure kind in the shortest time and launch debug runs

- A means to read the natural language executable verification plans described earlier, and annotate coverage to each feature providing feature-based views of coverage for analysis and reporting

vManager was deployed to manage regressions, enable interactive analysis of simulation results by individual block-owner, and create progress reports for use by the verification lead and management.

The vManager regression runner works by calling a user-script that launches a single simulation run. This required initial investment by one engineer to implement a basic single-run script, a simple means for each engineer to specify how to use it for each DUV and hooks to the vManager integration. This one-time effort paid off by enabling the use of the required verification management methodology for all DUVs on this and future projects. Figure 4 shows how vManager was used for verification management with the following use model:

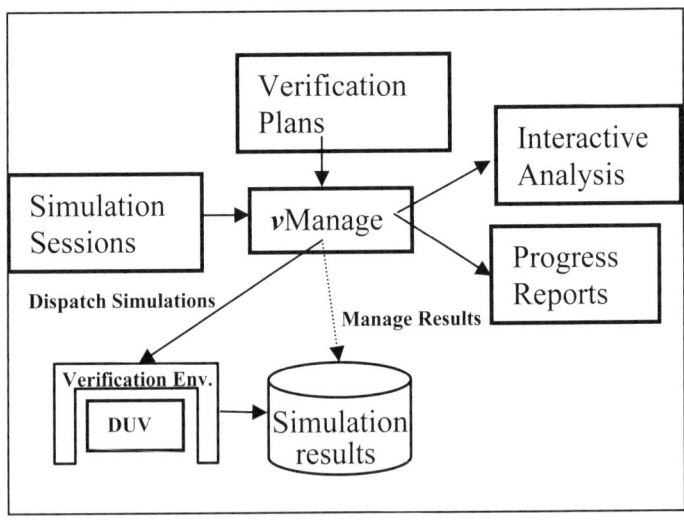

Figure 4 Regression management using vManager

Verification of a Next-Generation Single-Chip ATV and DTV ASIC   317

- Engineers created session descriptions that specified which simulations to run, vManager launched the simulations to the compute farm and managed the input and result files associated with each simulation
- Analysis of simulation failures was performed interactively in vManager
- The executable verification plans were read into vManager and used for interactive analysis of coverage for each feature and to create progress reports

Verification plans created in the planning process facilitated the tracking of progress at a very detailed level while organizing the results by feature to assess progress at a higher level of abstraction. This was achieved by annotating coverage metrics to each feature in the verification plan document, and using vManager to create feature-based views of the coverage for analysis and reporting. The snapshot of the example vPlan is Figure 5.

Section	Rel. Grade	Grade	Completion	Goal	Weight	At Least	Valid	Sampled	Items	Planned
4 Top	63%	63%	99%	100	1	1	12507	9100	2474	2489
4.1 Functional Requirements	63%	63%	99%	100	1	1	12507	9100	2474	2489
4.1.1 Reset and Initialization	74%	74%	100%	100	1	1	12137	8997	2430	2430
4.1.1.1 Register Accesses	74%	74%	100%	100	1	1	12137	8997	2430	2430
4.1.2 Interfaces	27%	27%	82%	100	1	1	280	57	9	11
4.1.2.1 Inter-block Interfaces	27%	27%	82%	100	1	1	280	57	9	11
4.1.2.1.1 P-bus Interface	53%	53%	71%	100	1	1	212	45	5	7
4.1.2.1.2 D-bus Interface	15%	15%	100%	100	1	1	34	6	2	2
4.1.2.1.3 L-bus Interface	15%	15%	100%	100	1	1	34	6	2	2

Figure 5 Regression results annotated to vPlan

The team lead played the progress tracking role using different stats, the most basic one being when all the regression suites were committed and included a sanity test case. Management got a summary report to assess which part of the chip required immediate attention, where to divert resources, etc. The results shared on the intranet in a standard format enabled everyone in different geographies to be in sync and have quick and easy access to results. On one occasion, for example, management noted the absence of block-level results for a specific block, and acceptable progress on another and diverted resources to focus on the block with less progress. The interactive analysis of results and reproducing the specific case with a waveform dump was quick and easy and thus, reduced the regression closure effort/time.

Finally, the answer to the question "Is the project finished yet?" was more reliable and was obtained without requiring excessive work by each engineer and the project lead each time an assessment of progress needed to be made.

Verification Environment Implementation
In this section we describe two aspects of the verification environment implementation that contributed to the results on this project, namely register verification and the reuse of verification environments.

Register Verification
Designs which are feature-rich and have several configuration modes rely on configuration registers to implement key features. There are 2600 32-bit in this chip, with a variety of access types. In addition to the more common Read/Write, Read-Only, Write-Only, and Clear-Read-Only types, a set of "Pending-Active" registers was implemented to hold configuration changes related to image processing in a pending state until the occurrence of an activation event (such as the start of blanking time between video frames). Each register needed to be verified for correct access type implementation of address, power-reset value, access-type, and function.

Automation was essential to manage the large set of registers and the subsets used in each DUV. A spec-driven process was implemented, whereby a custom tool extracts all the tables of the register spec word document, creates the database of registers and provides feedback on missing fields, incorrect address map, etc. The requirements for this custom tool were specified by the Design Verification (DV) team and implemented by the Software team. The output of the tool is various register definition files, one of which is in the format used by the Cadence register verification package which automates the creation of verification components for the register aspect of the verification environments, including shadow-register files, address-maps and register-access sequences.

Use of our custom extraction tool along with the register verification package enabled the team to:

- Synthesize the verification components related to register verification directly from the register specifications
- Extend the components to model the Pending-Active registers
- Create generic, self-checking stimulus sequences to verify power-reset values, address maps, and access types of the register set for each DUV
- Create reusable register access-sequences to configure the DUVs for various operations. These sequences were written at the level of abstraction of register fields

The automation and reuse afforded by this register verification scheme greatly reduced the effort associated with verification of registers in the block level, subsystem level and chip level. The register verification package was found to have enough hooks to easily add custom register access types.

Reuse of Verification Environments
When integrating two heterogeneous systems the important considerations in developing the chip-level testbench needs are maximizing the reuse of infrastructure developed at the subsystem level and detecting the first bug. Also important is building reusability into the new code for reuse in the bigger systems built later on.

The DTV subsystem was built using Specman Elite, the testbench automation tool from Cadence, and the ATV subsystem was built using a directed testbench with TCL and VHDL. Since the DTV subsystem environments were based on the e Reuse Methodology (eRM) from Cadence, they were far easier to reuse than the ATV subsystem environments that had been architected for top-level use only. So it was decided to build a Specman-based chip-level verification environment using eRM and System Verification Methodology (sVM) guidelines, and reuse the configuration sequences from the ATV subsystem verification environment.

There was valuable embedded knowledge in the TCL configuration sequences built for the ATV subsystem. A path for the TCL was built to generate configuration sequences as one of the configuration sequences at the chip-level. The TCL was modified to generate *e* code (eRM sequences) to configure the ATV subsystem at the top-level environment. The monitor developed for the ATV subsystem was hooked into a chip-level Specman monitor and scoreboard. This resulted in the chip-level testbench being composed quickly from the subsystem environments.

Chip-level tests were scoped to stress the end-to-end data paths, interrupt structure and system-level issues. The sample high-level data paths were playing audio, still picture on screen, playing video from an analog source, playing video from a digital source, Picture-In-Picture and overlaying OSD on the video screen, etc. These huge end-to-end data paths were further broken down into smaller paths that started or ended at memory. Small paths were debugged and stitched incrementally to make one full data path test. This sequence-based approach helped in the reuse of the block-level configuration sequences to build the chip-level data paths.

Table 4 Problem Domains Compared

Indicator	*Previous Chip*	*ATV–DTV Chip*	*Change*
Number of registers	1600	2600	+ 63%
Number of interfaces	7	12	5 NEW interfaces
Logic gates	Not Disclosed	Not Disclosed	DOUBLE
Regression size	500	1100	+120%

Results

The verification was completed on schedule! Tables 4–6 summarize some key attributes of the DUV which affected verification complexity, along with the results. The numbers are compared with those of the previous project completed a year ago. The verification strategy, tools, and well-defined reuse enabled to the completion of the verification within the scheduled tape-out date. Table 5 also identifies which factors contributed significantly to the efficiency (Key Enablers).

Table 5 Solution Domains Compared

Indicator	Previous Chip	ATV–DTV Chip	Key Enabler
Number of DUVs	30	24	Verification strategy
Team size	19	10	Verification strategy
Number of machines	20	16	vManager and LSF
Regression time	48 hours	60 hours	vManager and LSF
Regression closure (test suite complete to regression passing)	20 weeks	6 weeks	vManager

Table 6 Results Summarized

Indicator	Previous Chip	ATV–DTV Chip	Key Enabler
Spec-to-tape out time	18 months	10 months	Verification strategy, Reuse and vManager
Number of bugs found	230	170	Techniques, Specman and vManager
Number of bugs in FPGA	20	3	Strict CDV, More regressions, More reviews

Conclusions

Verification of digital TV enabler ASICs within the required schedules presents very interesting verification challenges. In this case study of the verification of a single-chip ATV–DTV ASIC discussed were the key areas that needed to be addressed, the approaches selected, and the results presented that showed the verification of a chip double the size of the previous chip was completed in approximately half the time with a 50% smaller team. Verification of this ASIC was completed on schedule with the allocated resources.

It was determined that the right balance of strategy, methodology, and tools will help address the verification challenges. Assessing verification challenges and developing a strategy is the first step. A sound verification planning and regression-management methodology implemented with tools that support an executable verification plan, regression-automation, and reporting was a key enabler. Verification environments that address reuse and pay special attention to register verification were also found to be key enablers of the success achieved on the project.

Future Work

The challenges associated with verifying Consumer Products will continue to intensify, and this requires verification teams to continuously improve processes and methodologies. Along with recounting successes, a review of a completed project must identify areas for future improvement. We will briefly describe three such areas, namely adherence to the reuse methodology, CPU-related verification methodology, and verification of the ATV subsystem.

Adherence to the Reuse Methodology

In previous sections we described how verification components based on the e Reuse Methodology were key to composing a chip-level environment from components created for the module-level verification effort. As the chip-level testbench was built, some components were found to be noncompliant with key aspects of the methodology that would enable reuse. This is understandable given a large team with varied experience, but the recoding required to correct these issues required redoing work at the module level, delayed the chip-level verification effort and diverted resources from the team's main goal of finding bugs.

To address this issue in the future the team plans to explore three areas: Additional code reviews to focus on reuse, automated generation of code that is correct by construction, and the use of static design analysis tools on the verification code to detect issues that would impede reuse.

CPU-Related Verification Methodology
The presence of an embedded CPU in a DUV presents several challenges. The CPU test code from the previous project had hard coded addresses that required numerous changes for the new address map. Considerable effort was spent to recode these tests as modular assembly code invoked by reusable sequences. This practice will be continued in future projects to keep this code reusable.

Another challenge experienced in this area was in reproducing in simulation bugs found when running the application software on silicon. The assembly code approach used in the simulation environment made this very difficult, and the team will explore the use of more advanced Hardware–Software verification techniques in the simulation area that would allow the use of C code rather than assembly code, along with the use of improved hardware-emulation techniques.

Verification of the ATV Subsystem
The ATV subsystem has evolved over several years and hence has a legacy verification environment that is based on a directed test rather than coverage-driven methodology, and does not have verification components that are reusable at higher levels of integration.

While the legacy environment has been adequate for previous generations of the product, the challenges experienced in verification when modifying and integrating this subsystem have shown that the verification approach needs to be updated for the current and future generations of the product. A particular area of concern was the porting of the environment for the new product which required one full-time engineer.

Accordingly, verification work on future projects will include creation of eRM-compliant verification components for new modules, recoding of existing verification components with advanced methodology where benefits are expected to be worth the investment, and interfacing between old and new verification components in environments where both are present.

Management IP: New Frontier Providing Value Enterprise-Wide

Steve Brown, Director of Marketing, Cadence Verification Division

Steve Brown is Director of Marketing for Enterprise Verification Process Automation at Cadence Design Systems. He is a 20-year veteran of the EDA verification industry and has held various engineering and marketing positions at Cadence, Verisity, Synopsys, and Mentor Graphics. He specializes in solving engineering, management, and marketing challenges that arise when new technology and products enter the market. He earned BSEE and MSEE degrees from Oregon State University and has studied marketing strategy at Harvard, Stanford, Kellogg, and Wharton.

Discussion around intellectual property in EDA has long focused on topics such as language, tool interoperability, and encryption. The domains of IP grew from manufacturing IP, to layout GDS-II, to designer RTL, and are rapidly entering the challenging world of verification. The health of the electronics industry is greatly improved by advances that facilitate IP reuse and exchange. There are changes underway that will accelerate and amplify the return company's realize from their investment in IP. Executives are beginning to realize that they themselves create valuable Intellectual property. The know-how, methods, and proven procedures they use to operate their teams effectively under increasing market pressure is IP that is extremely valuable. Let's take a closer look at how this form of IP will open up a whole new frontier of opportunities and define processes and automation that will increases project-level predictability and productivity.

I know managers creating anything as intelligent sounding as intellectual property sounds like an oxymoron. But think of how valuable it is to know how to estimate the productivity of an engineer, to assess the risks of various aspects of the system to be developed, and to monitor metrics trends throughout the life of a project. With these capabilities in hand management teams would be able to much better predict time to tape out and volume shipment. Imagine if that knowledge were captured in the form of data, in a system that automated the use of that data for planning, and providing data mining or analysis. The result would be a much more efficient running project with much higher quality. All because the management team had previous experiences captured in a format to reuse and leverage to make good solid management decisions.

For the sake of argument, let's take a quick look at an overly simplified example. Let's say you have an IP core or set of verification IP protocols that you've used successfully in previous designs. This IP has been tested and proven on the job. From a user's point of view that specific IP was a success – but from a manager's point of view that IP was not only a success, but now it is also an asset to leverage on future projects.

With experience comes various metrics that need to be gathered, stored, and managed. A manager's ability to estimate schedules, identify potential risks, and incorporate late stage design changes will all improve by leveraging prior experiences. Enterprise level software that captures, stores, and duplicates successful patterns for that specific IP model or verification IP are the first major step into the world of management IP. Management IP offers a foundation to plan and predict highly distributed resources needed on future projects.

To get to this foundation companies need to look at it from a different angle. IP needs to be categorized based on metrics captured while in use within the module or system. This means spending more time on evaluating each design and paying much more attention to the process, overall resources, and supporting methodology. Most crucial are resource productivity metrics – those leading and trailing indicators that are used to judge the project status. These data points are the beginnings of a complete management IP database.

This may sound simple but it's not – it takes a commitment from management to establish new processes and software solutions in a design flow.

There are many examples of IP data which needs to be captured. What if we could measure overall bug rates per engineer using a specific protocol? Or monitor the lines of code per engineer on average when using a specific IP module? How do you translate a system-level specification into lines of code that can be managed? What about change control and managing the "rippling affect" those changes have upon introduction to the project? All are very important data points that will greatly aid in your ability to design, schedule, budget, and predict your next project – the beginnings of a true "manageable" IP database.

Management IP may sound obvious, but the industry is showing clear evidence that the obvious is being overlooked. And if companies want to take full advantage of the wondrous time savings IP can offer, they will need to start capturing data, and develop best practices and skills at the management level. This will require working with companies that deliver process automation-based solutions that include managed solutions that address IP. When a premium is put on management IP enterprise-wide we'll see a much brighter outlook in the overall market, and your IP will no longer have to be alone.

Adelante VD3204x Core, Subsystem, and SoC Verification

Roger Witlox, Ronald Heijmans, and Chris Wieckardt
DSP-Innovation Center NXP Semiconductors

Roger Witlox joined Philips Research Laboratories in Eindhoven, The Netherlands in 1992, where he has been working on optical coherent communications systems and access networks Mr. Witlox was earlier involved in the development of analog laser temperature and current control system. In 2000, he joined the CTO organization at Philips Semiconductors, where he was responsible for development and support of an in-house verification tool. He has been responsible for functional verification methodologies for hardware IP and was a member of the Verification Technical Work Group of the SPIRIT consortium. In 2004, he joined the DSP Innovation Center and is currently focusing on DSP subsystems, both specification and verification.

Ronald Heijmans studied at the Hoge School Eindhoven and graduated in 1992 in the field of "Technical Computer Science." He started his career as a PCB designer at the Philips Research Laboratories. Later, Mr. Heijmans focused on DSP algorithm design and applications for multichannel audio and speech coding. In 1999, he became a verification engineer at ESTC Philips Semiconductors, where he focused on DSP core and subsystems. Currently, as a verification architect, Ronald is defining a new environment including new verification methodologies.

 Chris Wieckardt has been a Verification Engineer at Philips Semiconductors, Adelante Technologies, and NXP Semiconductors in Eindhoven, The Netherlands since 2000. Prior to Philips, Mr. Wieckardt was a digital Design Engineer at Océ Research and Development, Venlo, The Netherlands.

Abstract

It is impossible to verify the complete functionality of an IP at all levels (DSP core, subsystem, and SoC), therefore trade-offs between minimizing the number of used verification methodologies and having a quality product is a challenge. In this paper pros and cons of the verification methodology choices, made at the various levels, will be explained. The DUV is an embedded vector processor (Adelante™ VD3204x from NXP semiconductors). Furthermore some aspects of the verification process and the resulting verification plan will be discussed.

Introduction

The electronics industry trade press has been talking about the verification problem for SoC designs for a good 10 years now. The argument is simple and compelling: As chips have grown to the size and complexity of last generation's complete systems, verification becomes a challenge.

However, there has been less attention paid in the press to the problems of verifying IP cores, which are increasingly large and complex. In fact, there's an obvious but rarely stated corollary here: As chips grow to the size of systems, cores grow to the size of chips. Accordingly, the verification challenge for cores is also growing, and many of the same techniques adopted for SoCs are making their way into the development process for IP.

Perhaps nowhere is this more true than for processor cores, including general-purpose CPUs, floating-point units, and digital signal processors (DSPs). This article focuses specifically on DSPs, using a recent core development project at NXP Semiconductors (formerly Philips Semiconductors) as an example. We provide some background on the particular project, outline the challenges we faced, and describe the tools and techniques that we used to advance the verification process and improve the quality of our DSP core product.

Project Background

In mid-2004, our team embarked upon a two-year project to design and verify a completely new DSP with significant enhancements over previous cores. This core, which is now marketed as the Adelante™ VD3204x DSP, is built upon a Very Long Instruction Word (VLIW) architecture to support a significant degree of parallelism for both scalar and vector operations. Figure 1 shows the block-level architecture of the core.

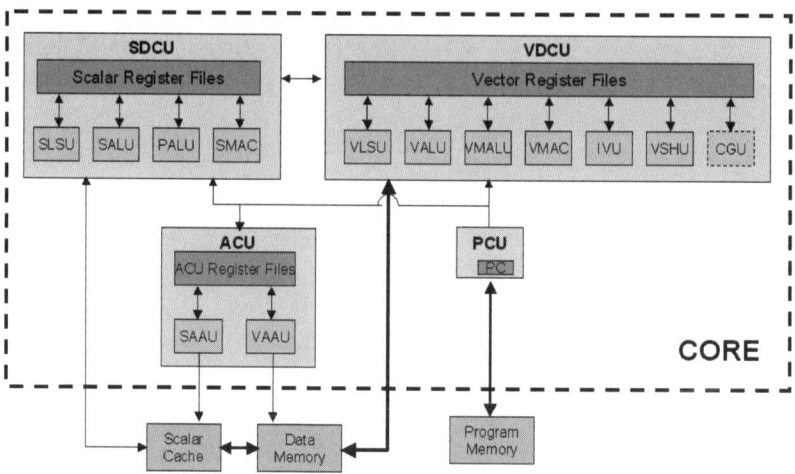

Figure 1 The Adelante VD32040 Core Contains Four Major Functional Units

The DSP core is divided into four main functional units:
- Program Control Unit (PCU)
- Address Computation Unit (ACU)
- Scalar Data Computation Unit (SDCU)
- Vector Data Computation Unit (VDCU)

The 256-bit instruction words, as is characteristic in VLIW architectures, can encode multiple scalar and vector operations at one time. As shown in Figure 1, the SDCU and VDCU have several sub-units to perform different types of operations, many of which can occur in parallel. Because of shared resources such as registers and buses, not all types of operations can be executed in parallel. Thus, there are complex rules about how the instruction words are constructed by a compiler. These rules had to be mimicked in the verification environment in order to test the DSP properly.

Figure 1 also shows an example configuration of three memories that reside outside the core:

- Program memory (4K × 256)
- Data memory (4K × 256)
- Four-way set-associative data cache (128 × 256)

These three memories form part of the DSP subsystem (DSS), which is shown in Figure 2. This subsystem includes a number of peripheral functions. These include support for emulation, tracing,

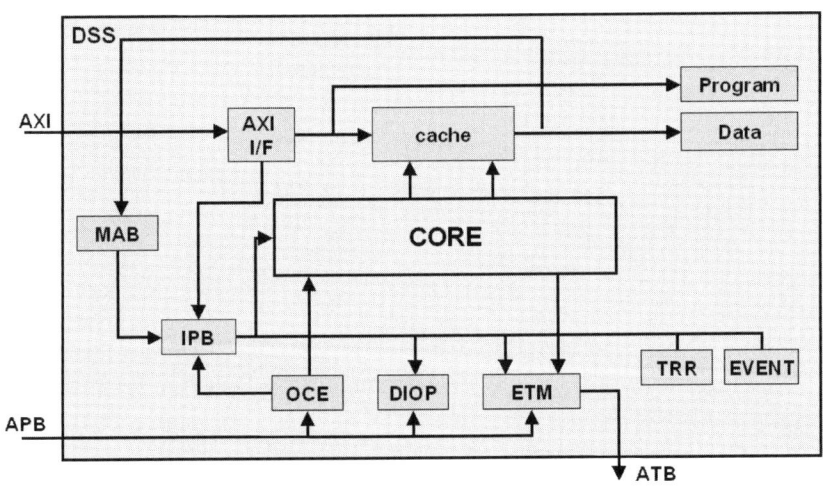

Figure 2 The VD32040 DSP Subsystem (DSS) Provides Memories, Peripheral Functions, and AMBA Interfaces

and multicore debugging, all of which run at the 300 MHz core clock rate, as well as several interfaces using the AMBA® 3 AXI™, AMBA Peripheral Bus (APB™), and AMBA Trace Bus (ATB™) on-chip bus protocols defined by ARM.

Verification Decisions

The nature of the VLIW architecture, with multiple operations in parallel running in a multistage execution pipeline, means that there are many corner-case conditions that must be exercised during the verification process. Of course, corner-case conditions lead inevitably to corner-case bugs, each of which must be detected and fixed before a design is released. In the case of the DSP core, we have many internal and external customers who depend upon receiving a robust, high-quality core so that they can concentrate on verifying the integration and their own logic, not on finding bugs in the core itself.

The VD3204x was a new core designed from scratch, significantly more complex than older IP, and so the project offered the opportunity for – and demanded – a more sophisticated verification approach. We decided to move beyond our traditional approach of using random-generated tests with a few functional coverage points.

Leveraging experience in NXP groups as well as in our own team, we chose a full plan-to-closure methodology using a coverage-driven, constrained random verification architecture. To accomplish this, we used Cadence's Incisive® Enterprise Specman® Elite test-testbench automation solution and Incisive Enterprise Manager verification management.

An important component of this methodology is the executable verification plan (vPlan), an online replacement for paper test plans that fosters early identification of verification and coverage goals, and enables reporting of progress against these goals throughout the course of the project. We found the vPlan to be useful both for verifying individual blocks and for integrating these blocks into a major subunit or a complete chip.

Figure 3 shows an actual report on some of the functional coverage points in the core. The top of the screen shot shows part of the hierarchy of features to be verified; the bottom shows the status of the functional coverage points associated with one specific feature (circled). The details of this screen shot are not important; the process is. We began by identifying specific features to be verified as part of our verification planning process, including corner cases that we wanted to exercise, and captured these in a Microsoft Word document using a template compatible with Enterprise Manager.

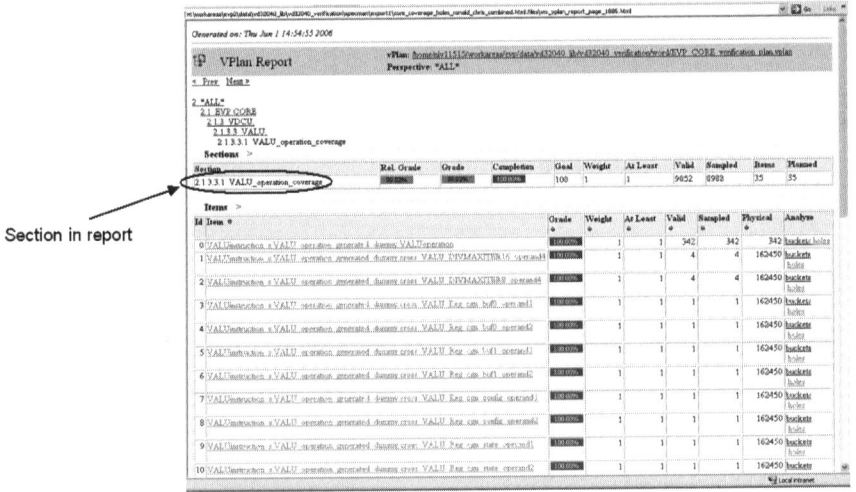

Figure 3 The Reported Coverage Metrics were Correlated Back to the Original Verification Plan

When we wrote the functional coverage code (in the *e* language) within the testbench, we added mapping information that identified the feature associated with each functional coverage point. (We could have added pointers to the coverage points in the vPlan instead, but we preferred to keep the original vPlan unchanged and link to it from our coverage code.) Throughout the verification process, we ran reports to show which points had been hit and which had not. Thus, our coverage metrics were always correlated back to our original vPlan, a critical part of a true coverage-driven methodology.

We also chose a multi-leveled verification strategy, specifically separating the verification of the core from the rest of the DSS. We made this decision primarily to avoid dependencies during the development process. We developed a BFM in *e* for the core that had sufficient functionality for DSS verification, which could occur in parallel with RTL coding and verification of the core itself.

We then entered a final verification phase that focused on the integration of the RTL core and the other DSS components, running actual software. Our customers perform similar tests once they have integrated the core and DSS into their final chips. The remainder of this chapter discusses the verification of the DSP core and the DSS.

DSP Core Verification

Prior to verification of the core as a whole, some of the RTL designers performed basic "sanity testing" on their individual blocks. They typically wrote some simple behavioral HDL code to stimulate the block inputs and examined waveforms of the block outputs to verify basic functionality. This approach was used primarily for interface blocks; the majority of the individual blocks were tested using a subset of the core test environment.

Our stand-alone verification of the complete core relied extensively on an instruction-set simulator (ISS) developed by our software development kit (SDK) team. This ISS was pipeline-execution-cycle-accurate in terms of all registers defined in the instruction-set architecture (ISA) for the DSP. Thus, at the end of every execution cycle, the state of these registers defined precisely what the corresponding registers in the RTL should also contain. Comparing the state of the ISS and RTL registers cycle-by-cycle was the single most important method used in core verification.

Figure 4 shows the complete core verification environment. We implemented all core tests as programs running in the DSP core. These programs were created by a test generator, written in *e*, which produced a series of instructions to try all the different DSP vector and scalar operations with different operand values. The constraints for test generation, also expressed in *e*, fell into two categories. Our "hard" constraints captured the fundamental rules about the DSP instruction set, such as which operations can be performed in parallel. As mentioned previously, shared resources within the SDCU and VDCU prohibited some types of subunit operations in parallel.

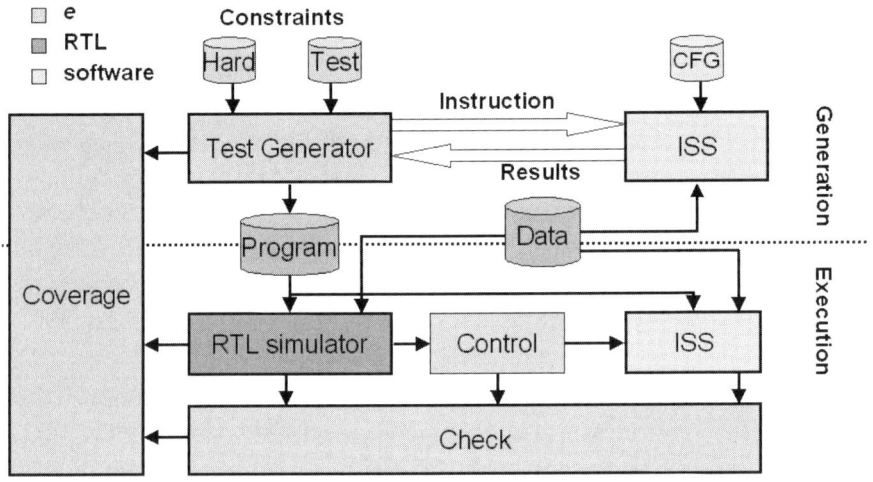

Figure 4 The Core Verification Flow Used Components Written in *e* to link the RTL and ISS Simulations

Our "test" constraints specified which specific subunits and which particular operations within these subunits we wanted to test. Given the large number of operation combinations to be tested and the limited size of the DSP's program memory, we had numerous small tests rather than a few very long tests that tried to verify the complete core. Having multiple tests also allowed us to execute in parallel on a simulation server farm, reducing the total time for each regression run. The output of the test generation phase was a program image that was loaded directly into the DSP's program memory and executed in RTL simulation.

Although our hard constraints captured most of the architectural dependencies, there were more subtle resource conflicts that could develop between multiple instructions in different stages of the execution pipeline. We didn't attempt to capture all of these rules with constraints, but instead leveraged the ISS and its detailed architectural model of the instruction set. In addition to computing the expected results for each instruction, the ISS checked the validity of the instruction against other instructions in the pipeline. If it found any resource conflicts, it reported an error condition to the test generator. The test generator had the ability to "back up" and reissue instructions to avoid these conflicts. Only conflict-free instructions were actually included in the final test program.

When each test program was executed in the core verification environment, the ISS ran the program in parallel with the DSP RTL. Because these two representations were aligned, we included e check code that compared the state of all ISA-defined registers at the end of each pipeline execution cycle. This primary bug-finding approach proved very effective, although it meant that we had to wait until the ISS was complete and fairly robust before starting RTL core verification. In the process of verifying the RTL against the ISS we found some bugs in the ISS as well, so the verification engineers worked closely with the architects to resolve any differences.

Given our focus on coverage-driven verification, we used Specman Elite to collect detailed coverage metrics for each test and to merge the results together to yield an overall view of our verification progress. All the functional coverage points were expressed by the verification engineers as part of coding the core verification environment, but other members of the team contributed significantly to the identification of important coverage points by contributing features to the vPlan. For example, our architects had numerous ideas about critical corner cases that had to be exercised in order to declare verification complete.

Our designers also participated in the coverage process by providing vPlan features. As they designed the microarchitecture of their blocks and coded the RTL, they were strongly encouraged to think of interesting states that should be checked by coverage points. This is one case in which the executable verification plan really helped; the designers could simply document their corner-case features in the vPlan without worrying about how and when the corresponding coverage points would be coded by the verification engineers. This resulted in a much more comprehensive use of functional coverage than had been the case on our previous DSP projects.

In addition to functional coverage points, we used traditional code coverage metrics to help identify portions of the RTL code that were not being sufficiently tested. We used our simulator's native code coverage capabilities to collect results from each simulation test and to merge these results together. We did not link the code coverage goals or results into the VPlan, although we did write some scripts to combine the Specman Elite functional coverage reports with our simulator's code coverage reports to produce a single top-level coverage view.

Before moving on to the DSS verification, it is important to note that we focused the stand-alone core verification only on proper execution of the instructions. Certain other aspects of the DSP operation, such as interrupt handling, were tested in the DSS verification process and confirmed during integration verification. The next section discusses the verification of the DSS components and the overall subsystem in more detail.

DSP Subsystem Verification

Before verifying the entire subsystem, we verified several of the DSS RTL components with stand-alone testbenches. We found two advantages to stand-alone verification of the cache logic, the OnChip Emulation (OCE) component, and the Embedded Trace Module (ETM). First, it was easier for us to exercise corner cases from a separate testbench rather than within the entire DSS. For example, we wanted to fully exercise the cache's snoop-port arbiter, which required precise control overcache accesses.

In addition, we found that parallel testing of components avoided dependencies that would have delayed verification. For example, the AXI interface was under development at the same time that we were verifying the cache logic. If we had waited for a DSS environment to verify the cache, it could have happened only after the AXI design was complete.

We used a classic *e*-based testbench, running sequences of traffic and using a scoreboard to keep track of results, to verify the three stand-alone DSS blocks. Figure 5 shows the testbench for the OCE, which bridged to the core's Internal Peripheral Bus (IPB) and a DSS debug bus. The OCE provided access to both the core's ISA-defined registers and the DSS peripherals via the external APB for the purposes of multicore debugging. This testbench used several *e*VCs, including a "shareware" APB *e*VC and several written by our verification engineers.

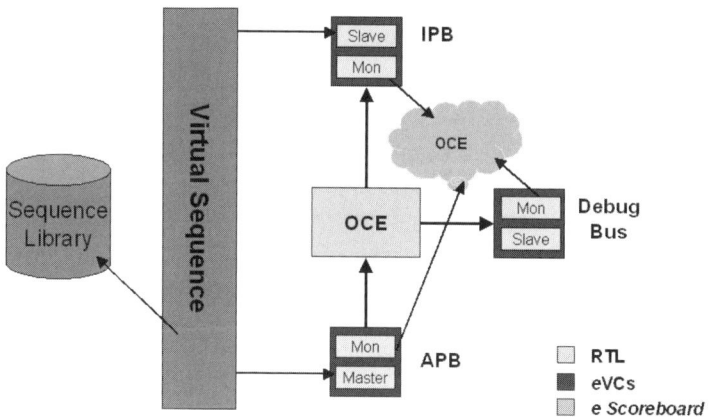

Figure 5 The Stand-Alone Testbenches for the OCE Used Interface *e*VCs to Generate Bus Traffic

In parallel with this component-level verification process, the verification engineers developed the BFM for the core. In addition to allowing an earlier start to DSS verification, the BFM provided more fine-grained control of constrained random stimulus and simulated faster than the full RTL implementation. Figure 6 shows the major functions of the core BFM, as well as the entire DSS verification environment. As was the case with the component testbenches, this environment used *e*VCs and scoreboards to generate traffic and track results.

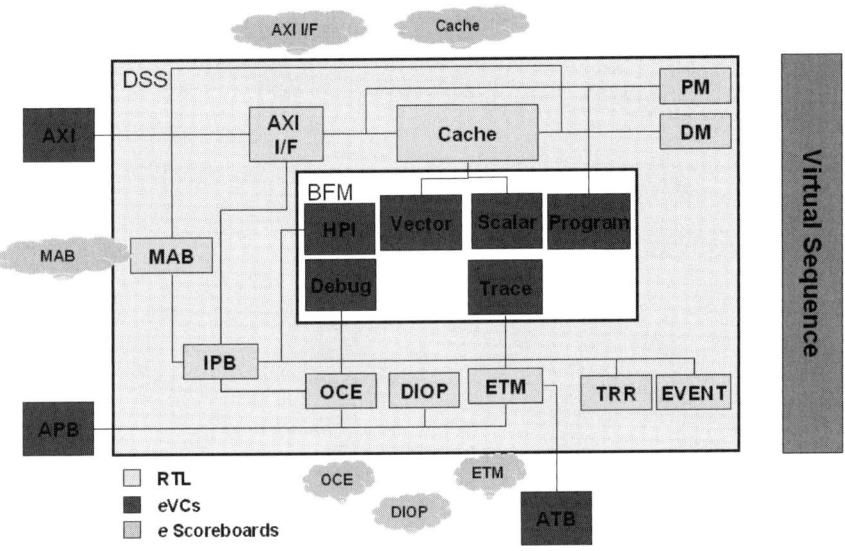

Figure 6 Most Verification of the DSS Used a BFM of the DSP Core

One of the advantages of a constrained random, coverage-driven environment built using *e*'s object-oriented capabilities is a high degree of verification reuse. In fact, during full DSS verification we were able to reuse the component-level *e*VCs, scoreboards, and sequences. We were also able to use a commercial AXI *e*VC as well as the shareware APB *e*VC, allowing us to leverage previous verification work performed by others in the industry.

Coverage metrics were every bit as important for DSS verification as they were for DSP core verification. We created a DSS vPlan with a detailed feature list based upon recommendations from the architects, designers, and verification engineers, and linked to these features when writing the functional coverage code. When running tests, we collected code coverage metrics for all the RTL DSS components, and we again used Specman Elite to collect and merge functional coverage results from the core BFM, the *e*VCs, the scoreboards, and other portions of the verification environment.

SoC-Level Verification

After the verification of the core and subsystem, the next step in a verification process is verifying the integration of the subsystem in an SoC. Normally this is a task of the integrator but the methodology used enables an IP provider to deliver integration tests together with the IP.

The methodology used is C-based with the focus on interconnect and interoperability verification and assumes the IP itself is functionally correct. One of the advantages of using a C-based approach is the reusability across platforms (e.g., RTL simulation, prototyping, or final silicon). This reuse requirement is one of the main reasons why the verification methodology used at SoC level often differs from the ones used at core or subsystem level.

Figure 7 shows an example of verifying interconnect and interoperability of the trace part of an SoC using this methodology. A C-program, running on the VD3204x, configures the other trace components (funnel, ETB) in the system, using the AXI interface. In the next step trace data will be generated and stored in the embedded trace buffer (ETB), using the ATB busses. Finally the VD3204x can read the content from the ETB and compare this content with the expected value.

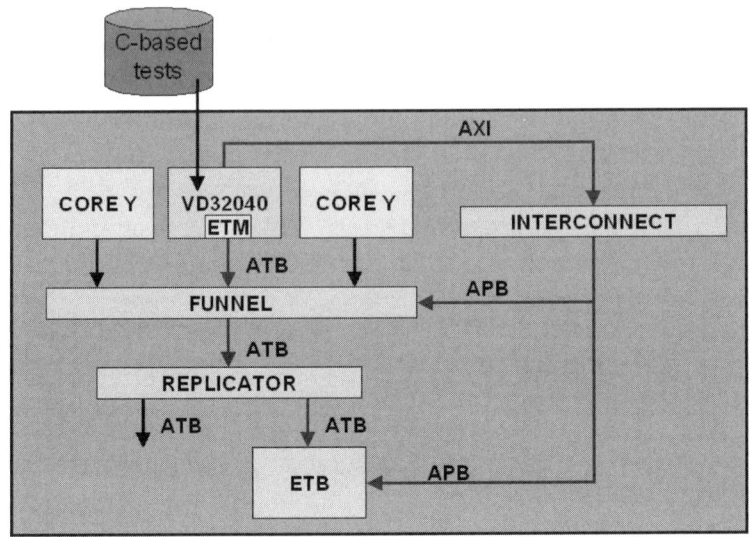

Figure 7 SoC-Level Verification was Performed Using C-Based Tests

Results and Future Work

Figure 8 summarizes the specific techniques used in the different layers within our overall verification strategy. The three levels discussed in detail in this chapter have several common themes: extensive use of constrained random stimulus, Specman Elite running sophisticated *e*-based verification environments, Enterprise Manager's executable verification plans, and reliance on coverage metrics to gauge verification progress. In addition to its planning capabilities, we found Enterprise Manager's ability to automate regression runs and report results in a concise manner very helpful.

	Core Level		Sub-System Level			SoC Level
	Core Component	Core	DSS Component	DSS without Core	DSS with Core	DSS Integration
Checks	Eye-ball Waveform	RTL vs. Simulator	Scoreboard	Scoreboard	Directed Test	Directed Test
Owner	Design Engineer	Verification Engineer	Verification Engineer	Verification Engineer	Verification Engineer	Customer
Stimuli	Behaviour RTL	Random Program	Random Traffic	Random TSaffic	Software Based	Software Based
Focus of this article?	✗	✓	✓	✓	✗	✗

Figure 8 Verification of the DSP Core and its Subsystem Used Different Techniques at Different Levels

Of course, all the advanced features of our methodology required a nontrivial investment. The setup and execution of the verification environment required about three engineer-years, nearly as much as the four engineer-years required to write the 60000 lines of RTL code in the core. This investment was clearly worthwhile; the Adelante VD3204x DSP core has already been used in successful SoC-design products, with the first tape-outs imminent, and no major problems have been reported by any internal or external customers.

We are very pleased that we met our goal of improved core quality, and we are quite certain that we will continue to use our proven methodology on derivatives of the VD3204x. We have a few ideas for enhancing our methodology, including using assertions both for more precise bug detection and for designers to express corner-case coverage points directly within their RTL code. We are certain that we will be able to use the experiences on this project as a baseline for high-quality design and verification on future projects involving both cores and complete SoCs.

SystemC-based Virtual SoC: An Integrated System-Level and Block-Level Verification Approach from Simulation to Coemulation

Laurent Ducousso, Frank Ghennassia, and Joseph Bulone
ST Microelectronics

Dr. Laurent Ducousso has over 20 years of experience in digital design and verification. In 1994, Dr. Ducousso joined STMicroelectronics as the verification expert on CPU, microcontroller, and DSP projects. Since 2000, he has managed the Home Entertainment Group verification team. Prior to STMicroelectronics, he was engaged in mainframe CPU development at Bull S.A for 8 years. Laurent holds a Ph.D. in Computer Sciences from Paris, France.

Frank Ghenassia is Director of the System Platforms Group in the HPC (Home, Portable, and Communication) sector at STMicroelectronics. Mr. Ghenassia focuses on IP/SOC verification, architecture definition, platform automation, and embedded software development based on high-level modeling approaches. He joined STMicroelectronics in 1995 and has worked on OS development, software debuggers, and system-to-RTL design flow activity. Mr. Ghenassia received his M.S. in Electrical Engineering in Israel.

Dr. Joseph Bulone manages a team that provides central services in hardware emulation to STMicroelectronics divisions. Joseph defines and provides hardware-accelerated platforms for IP/SoC verification and software development. He joined the Central R&D division of STMicroelectronics in 1989, and was initially involved in the design of ATM chips. He began working on hardware emulation in 1993. He has been in charge of video chip validation, and hardware software codesign. He holds a Ph.D. in microelectronics from the Institut National Polytechnique de Grenoble, France.

Abstract

ST faced two daunting challenges for their next generation product (1) to provide an advanced and fast platform for s/w development, including ISS and hardware models described in abstraction level, running at a minimum targeted rate of 1 MHz in the simulation environment and (2) to integrate the system level and block-level verification environments for a large RTL design with a significant firmware component.

The Transaction-Level Modeling (TLM) capabilities of SystemC were used to deliver a Virtual SoC and helped to resolve challenge number (1). Though the TLM behavior was modeled with more abstraction, there was enough accuracy for the software developers to be able to debug their SoC design while running at 1 MHz. Having this platform available early in the process enabled software engineers to begin developing the embedded software for the application. Not only did this bring in the overall project timescales, but also the exceptionally close cooperation between the software and hardware teams in the early phases of the project led to the detection of significant bugs in the hardware specification of the design. Because these bugs were found early, they were relatively cheap to fix, and contributed to save a respin of the chip.

The Virtual SoC was extended to provide a block-level Verification environment for a Low-Cost MPEG2 and more recently MPEG4 design using Incisive and SystemC. Reusing the system-level environment in this way means that the tests (using images 80 × 96 pixels) and test harness do not have to be reimplemented in a new language and tool set. In order to speed up the RTL verification regression and run full size conformal tests (with 1920 × 1080 and 720 × 1080 pixel images), Transaction-Based Acceleration (TBA) and emulation have completed the validation process.

In order to accelerate the regression test of the IP, virtual SoC environment used Cadence Incisive and the Palladium for signal-based acceleration, by reusing SystemC high level of abstraction for the testbench portion (simulations went from 300 Hz on Incisive up to 10 KHz using signal-based acceleration). This performance will improve further in the future, using TBA methodology. Incisive capabilities included mixed language SystemC/RTL kernel, SimVision for debugging and performance analysis thanks to TxE.[2] SysProbe methodology, dedicated to verify RTL performance and functionality, was then built on top of TxE.

Introduction: Verification and Validation Challenges

Because of the increasing complexity of set-top-box chips, the Verification team decided to follow SystemC/TLM methodology. This allowed SW teams to initiate their SW development early in the design flow and provide an advanced and fast cosimulation platform for s/w development. This included ISS models, running at a minimum targeted rate of 1 MHz in the simulation environment without the use of hardware accelerators. Figure 1 illustrates the SoC TLM flow compared to the old flow. Section 2 describes this Virtual SoC platform for 3 usages: SW development, HW verification and architecture exploration, and analysis. Section 3 will present ST verification process for RTL at block and platform levels for set-top-box chips.

---

[2] Transaction Explorer tool from Cadence.

In order to complete full verification regression tests (with real image sizes), the Virtual SoC platform was extended to include simulation acceleration. This approach makes use of the Transaction-Based Verification (TBV) methodology, which enables the ability to mix SystemC testbench with RTL emulated on the Palladium[3] H/W emulator. This will be described in Section 4. Section 5 will summarize the benefits of the Virtual SoC platform and will comment the next steps of ST TLM methodology.

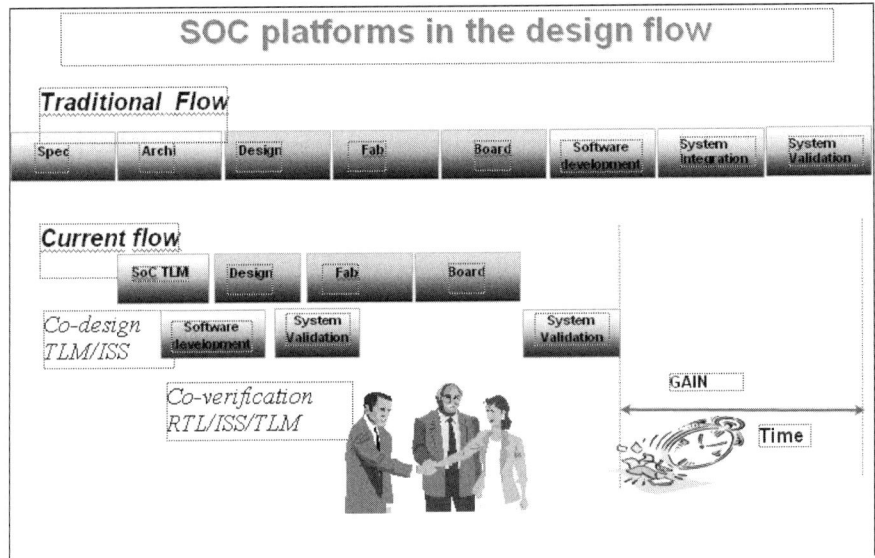

Figure 1 Comparison Between Traditional and SystemC/TLM Flow

Virtual SoC TLM Platform

Transaction-Level Modeling (TLM) was pushed by industry and research institutes through OSCI to respond to the following tasks:

- Embedded software development
- Functional verification
- Architecture analysis and exploration
- HW/SW coverification, HW validation

---

[3] Cadence H/W accelerator & emulator.

An Integrated System-Level and Block-Level Verification   349

TLM infrastructure was developed to support modeling communication structures at three abstraction levels, i.e., Programmer's View (PV), Programmer's View with Timing (PVT), and Cycle Accurate (CA), leaving it up to the user to compromise between simulation speed and accuracy. ST has played a major role in the OSCI-TLM working group and deployed TLM methodology on multiple projects.

The virtual SoC TLM platform was developed at the PV level from Specifications in order to offer fast simulations for the next phases. This platform has been used as reference model and enabled concurrent SW and HW engineering and close cooperation in early phases of the project. This process led to the detection of significant bugs early in the hardware specification. Because these bugs were found early, they were relatively cheap to fix, and contributed to save a respin of the chip.

SW engineers could start development before having the board. As example, this was done on Graphic Engine Blitter and MPEG2 projects; the driver was developed before having a board, that led to 6 months time gain in comparison with traditional flow (as pointed out in Figure 1).

HW verification group employed TLM platform because, though more abstract, it accurately modeled the bit-level behavior of the SoC while running at 1 MHz (this was achieved on MPEG4 decoder project). This will be fully described in the following section.

Another domain of utility is architecture exploration and analysis (Figure 2). The SoC TLM platform, when refined with timing information, can provide relevant information on bus bandwidth, peripheral accesses, interrupt latencies, memory conflicts, and latency to the architects. The SysProbe methodology was built at ST using the flexible transaction recording, viewing, and analysis capabilities of Cadence's SimVision and TXE. SysProbe could record the transactions generated by proprietary architectural models. It was also used for functional and timed validation. By calibrating TLM with back-annotated data [1] it was also possible to radio the same transactions generated by either TLM models or corresponding

RTL models and to compare the results using the environment provided by Cadence's TXE. This technique was used to verify the performance of the RTL.

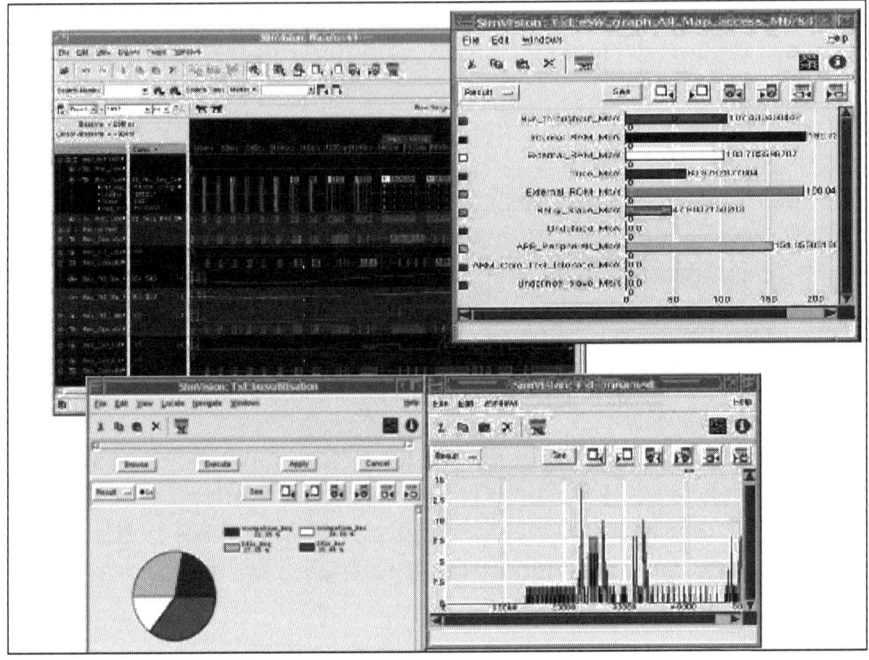

Figure 2 SimVision/TxE/SysProbe Analysis Environment

Functional Verification: Cosimulation TLM and RTL

The Virtual SoC TLM platform (integrating bus, memories, CPU) was also reused for functional verification at block level and platform level using Incisive and SystemC. This provided the ability to achieve fast simulations. As an example, for MPEG4 decoder, full RTL (including complete testbench) would take 6/7 hours. This was reduced to few minutes with TLM backbone (tests used images 80 × 96 pixels).

An Integrated System-Level and Block-Level Verification 351

Reusing the system-level environment at block and platform views means that the tests and test harness do not have to be reimplemented in a new language and tool set. A three-step approach was used to achieve block- and platform-level verification:

- First step is to verify the block level. This step used the TLM models of the low-cost MPEG4 already developed for the virtual SoC. This implies that the RTL blocks can be verified stand-alone before integrating with the rest of the RTL. It also means that function tests used for system-level verification can be reused for block-level verification.
- The Second step involves connecting the RTL DUT to the TLM testbench through BFM, also known as transactors. These transactors use the SystemC Verification Library (SCV) to shape the timing characteristics of the traffic across the bus, and the transaction recording and viewing capabilities of Incisive to verify the performance and functionality of the Design Under Test (DUT).
- The third step is to use the fast mixed language simulation and debugging facilities of Incisive to verify the full RTL design, by connecting the RTL description of the DUT to the SystemC-based Virtual SoC Verification environment described in previous sections.

Writing tests at the transaction level means that the tests can be used at both system and block level. But to do effective block-level verification, we need to stress the DUT by shaping the timing characteristics of the data. ST uses SCV, the OSCI verification library supported by Incisive, to randomize the timing characteristics of the bus traffic. For example, we can allow the length of a burst write to vary between a minimum and maximum number of clock cycles, or we can specify the gap between one burst and the next. Randomizing traffic characteristics in this way can trap costly bugs that the block designer may not have been able to test for.

SysProbe methodology is used together with SimVision as a powerful transaction-level visualization tool. By visually looking at transactions rather than individual signals in a waveform viewer, functional bugs can be identified and tracked down a lot quicker, and more efficiently. Once the problem is identified, the verification engineer can switch to the signal level to work out how to fix the bug.

Validation: Coemulation TLM-Palladium

The Validation process is incomplete without testing real condition input to the design. Once the design was converted from SystemC to RTL, simulation performance was reduced. In the case of the MPEG4 design, full image sizes are 1920×1080 and 720× 480. These image sizes could not be fully tested during simulation.

The traditional approach at CMG group has been to wait until fully synthesizable and complete RTL is available, and then use the Palladium emulator to test full RTL implementation. The TBV methodology was adopted in order to get a head start for full-chip verification, early in the development effort. This alleviated the need to wait for the availability of complete RTL (including testbench). This methodology allows the design teams to reuse the SystemC testbench that was used in the first and second phase, while the design is being converted into RTL. The performance gain allows the teams to run long tests and continue to validate their system, while the DUT is getting its final RTL representation.

The Incisive TBA solution (as illustrated in Figure 3), which is based on the standard coemulation modeling interface (SCE-MI), enhances simulation acceleration performance of the Palladium system by reducing communication between the testbench running on the workstation and the DUT in the emulation system. Productivity features include support of variable-length messages, a faster streaming mode, transaction recording capabilities, and support of both timed and untimed testbench components. This solution enables full congruency with the Incisive unified simulator to shorten bring-up time and assure reusability of the testbench and the verification IP models. ST hopes to reach 100 kHz in comparison of 10 kHz at signal-based acceleration for a full image format SD.

An Integrated System-Level and Block-Level Verification

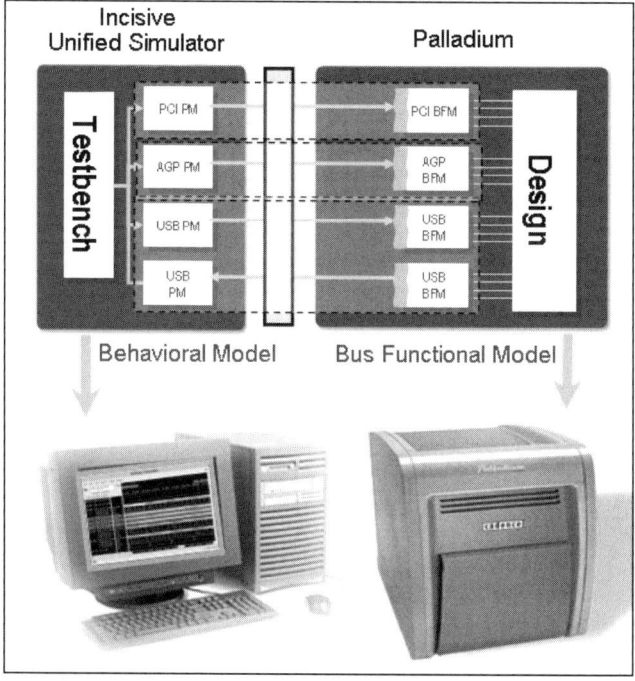

Figure 3 Transaction-Based Acceleration

Conclusion and Future Developments

In summary, the virtual SoC TLM platform made early SW development possible – up to 6 months earlier than the traditional approach. The Validation process was also pulled in by approximately 3 months – by utilizing the TBV methodology. All this was made possible by using SystemC/Incisive environment, which also provided the ability to maintain same debugging and transactional environment for both RTL and System-level verification.

The effort to create and update more and more TLM models into the portfolio is an ongoing process at ST for efficient System-Level Verification. TBA methodology is really beneficial if one can enhance the speed of coemulation. While this approach has already proven beneficial at ST, there is continued joint effort to improve this methodology and technique.

To further enhance the efficiency and throughput of the Verification effort, Cadence Assertion-Based Verification (ABV) is also being investigated, and will potentially be used on future projects.

1. Clouard, A., Mastrorocco, G., Carbognani, F., Perrin, A. und Ghenassia, F. (2002). Towards Bridging the Gap between SoC Transactional and Cycle-Accurate Levels. In Proc. of Design, Automation and Test in Europe – Design Forum (DATE'02), Paris, France. IEEE CS Press, Los Alamitos)

Is Your System-Level Project Benefiting from Collaboration or Headed to Chaos?

Steve Brown

Steve Brown is Director of Marketing for Enterprise Verification Process Automation at Cadence Design Systems. He is a 20-year veteran of the EDA verification industry and has held various engineering and marketing positions at Cadence, Verisity, Synopsys, and Mentor Graphics. He specializes in solving engineering, management, and marketing challenges that arise when new technology and products enter the market. He earned BSEE and MSEE degrees from Oregon State University and has studied marketing strategy at Harvard, Stanford, Kellogg, and Wharton.

In order to parallelize project operations and meet aggressive schedules, system project teams designing both hardware and embedded software must address the need for much higher frequency of interactions. Without better forms of communication, automation enhancements, and verification engines that are powerful and flexible, attempts to parallelize flows will result in chaos and project paralysis.

Industry trends in electronics are resulting in design and verification schedules becoming more compact and complex. The use of SW to implement more functionality provides flexibility, but also compounds the difficulty in completing verification because of the need to do more HW/SW coverification before silicon is available. The emergence of effective HW/SW coverification solutions alleviates this pressure technologically, but it creates an unanticipated burden for many project teams: frequent daily interactions between HW and SW teams as they converge on closure independently and collectively.

In the world of software design, engineers typically try to steer clear of the complex hardware verification process. Historically they have presumed a stable silicon platform upon which to run and test their drivers and applications. In some cases this results in finding hardware bugs late. But the benefit is that most of the busy-work of designing the hardware has been completed. As more of the hardware and software is coverified, the more the software team is exposed to the noisy, tumultuous process of reaching closure of the hardware. While in theory parallelizing hardware and software is an overall gain for the project, the team must address the significant increase in details that will be discovered and resolved while exposing the software team to the hardware development. Without an effective plan to handle the volume of issues, the project will stall and all parallelization benefits will be lost.

Communication Barriers Torn Down

When it comes to system-wide applications the "communication gap" is just not acceptable any longer. Hardware design and verification has matured to a point where we regularly see predictable schedule success when applying verification process automation. System verification projects must leverage these same approaches for HW/SW coverification and system-level closure. With more upfront planning between the hardware and software teams, communication and understanding of what the other is doing will bring huge benefits. The result of planning is documented intent, and measures of success using coverage data like code converage, assertion coverage, and functional coverage, made visible across the team. Assumptions that each team needs to make based on the initial system specification will be regularly reviewed, check-points will be established continually checking for system-level bugs, and a broad-based agreement on how the system should behave to reach full system-level closure will be agreed upon. The metrics built into the plan will apply to the necessary check points of the hardware and embedded software and be tracked and managed together in a system-wide management solutions leading to prioritization and escalation of problems and priorities if needed.

The use of metrics is what allows introduction of automation. This automation applies to the individual verification tasks: dispatching jobs, analyzing results, and debugging failures. The individual benefits of automation are only incremental in their impact on the overall project. Most important is the automation of handling the project-wide analysis, tracking, and decision making. These metrics make it easy for a centralized view of either the hardware or the software, and identification of critical problems to address first. Communication will become much more instant and less disruptive as the reporting and management tools take care of most of the work. Updating of the status will be more frequent with less overhead. Perhaps the most important improvement in the project team's development cycle will be the ability to adapt and react to changes. This clarity enables the software and hardware teams to work independently, while still rapidly uncovering and resolving critical system-level issues. Adjustments in resource focus and testing needs will adapt automatically based on the changes in the initial specification and how they flow through the rest of the plans.

Without applying these important and proven concepts of planning, metrics, and automation into your hardware and software development process you are only setting yourself up for project, level paralysis. Hardware and software verification have to be parallelized within the hardware and software domains to keep up and shorten schedules with limited resources, engines, and skill levels. The metric-driven approach itself will comfortably bring together the hardware and software teams by capturing common goals and check points of operation. High-performance models and engines, quality verification IP, coverage-driven verification approaches, and debugging across both software and hardware domains will become increasingly essential. In addition, this entire process will need to be managed at a project level from the block to chip and full system level. Without this collaboration I'm afraid we'll be headed for system-level chaos – which is exactly what we need to avoid.

Index

A

Acceleration, 75
Alfonso Íñiguez, xxiii
Architect
   design, 36
   verification, 35
Architectural Exploration, 58
Architectural Verification, 165
ARM926, 205
assertion, 46
Assertion based verification, 167
Assertion Based Verification, 63
Assertions, 44

B

black-box, 92
Block Verification, 45
brainstorming, 17, 85
Brainstorming, 87
Branch, 130
Bus Traffic, 120

C

cannot, xiii
cereal, 81
Chip Level Verification, 48
Chris Wieckardt, xxvi
Cloning, 141
closure
   tracking, 6
Code Coverage, 117, 160, 192
constrained random stimulus, 11
Constrained Random Testing,
   188, 194
Cost of Failure, 5

Coverage, 161
   Assertion, 161
   Checker, 161, 195
   Code, 196
Co-Verification, 197
Crisis, 4
CVS, 129

D

death by status meeting, 13
Debug, 141, 148
   hypothetical, 148
Debug coordinator, 39
Debug triage, 144
Design Engineer, 38
Design Manager, 34
design quality, 123
Design/System Architect, 36
Directed Testing, 187

E

e, 64
Egidio Pescari, xxii
Emulation, 75
engineer
   design, 96
engineers, xv
   design, xv
   entry level, xv
   verification, xv
environment definition file, 132
Euclid, xix
eVC, 203
executable plan, 55
executable verification plan, 18
execution, 15

implementation, 21
   verification, 21
Executive, 31
Executives, xv

F

FASB, 7
firmware, 63
formal, 46, 65
Formal, 169
formal verification, 65
Frank Ghenassia, xxvii
functional coverage, 71
Functional coverage, 43
Functional Coverage, 118, 157, 192

G

Gabriele Zarri, xxi
Generic Software Adapter, 210
Gibberish, 190
goal, 103

H

Hamilton Carter, xi

I

implementation execution, 21
Incisive Enterprise Manager, 25
integration verification, 98, 101
Integration Verification, 47
Iraklis Diamantidis, xxii
ISO, 7

J

Jason Andrews, xxi
Joseph Bulone, xxvii, 346

L

Laurent Ducousso, xxvii

M

Machine utilization, 125
managers, xv
   design, xv
   verification, xv
Mario Andretti, 3
Marketing, 33
Maslow, 9
Mediterranean Sea, xix
Metric-driven Process, 6
Metric-driven Process Automation Tools, 52
metrics, 18
Metrics
   too many, 125
milestone, 104
milestones, 103
Mixed Signal Verification, 73
moderator, 89
Monia Chiavacci, xxi

P

Paul Carzola, xxiv
planning, 15
Planning, 86
PLANNING, 240
planning session, 17, 84, 175, 185, 289
Planning Session, 89
Port, 214
postmortem, 20
process metrics, 143
product intent, 42, 58, 82
PSL, 64
Pythagoras, xx

R

regression, 23
   bring-up, 116
   design quality, 121
   integration, 119
   revision management, 115, 123

Index

Regression, 113
Regression Management, 113, 114
Regressions coordinator, 39
response, 16
Response, 26
Revision, 115, 129
   Tag, 134
Revision Control System, 129
Revision Tag, 131
Roger Witlox, xxvi
roles, 31
Ronald Heijmans, xxvi

S

sequences, 230
Sequences, 217
Shankar Hemmady, xi
Simulation, 70, 181
`Snooping`, 158
Software utilization, 125
Steve Brown, xxv
strategic metric, 29
Stubs, 216
Stylianos Diamantidis, xxii
Susan Peterson, xxiv
SVA, 65
System Level Verification, 48
SystemC, 163
SystemVerilog, 45

T

tactical metric, 29
Test-Based Coverage, 161
testbench, 70

Thanasis Oikonomou, xxiii
track
   assertions, 67
transaction level modeling, 43

U

universal coverage database, 108

V

vComponents, 230
verification, 48
   integration, 47
   system level, 48
Verification, 3
   block, 45
Verification
   Architect/Methodologist, 35
verification engine execution, 21
verification engineer, 92
Verification Engineer, 37
Verification Hierarchy of Needs, 9
Verification Manager, 34
Verification Planning, 86, 201
Verilog, 45, 64, 181
VHDL, 45, 64, 181
view, 102
Visibility, 136
vProbes, 231

W

Waveform, 146
white-box, 100

Printed in the United States of America.